Occurrence and Distribution of Pesticides in Surface Waters of the Hood River Basin, Oregon, 1999-2009

By Whitney B. Temple and Henry M. Johnson

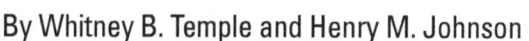

Prepared in cooperation with the Confederated Tribes of Warm Springs

Scientific Investigations Report 2011-5082

U.S. Department of the Interior
U.S. Geological Survey

U.S. Department of the Interior
KEN SALAZAR, Secretary

U.S. Geological Survey
Marcia K. McNutt, Director

U.S. Geological Survey, Reston, Virginia: 2011

For more information on the USGS—the Federal source for science about the Earth, its natural and living resources, natural hazards, and the environment, visit http://www.usgs.gov or call 1–888–ASK–USGS.

For an overview of USGS information products, including maps, imagery, and publications, visit http://www.usgs.gov/pubprod

To order this and other USGS information products, visit http://store.usgs.gov

Suggested citation:
Temple, W.B., and Johnson, H.M., 2011, Occurrence and distribution of pesticides in surface waters of the Hood River basin, Oregon, 1999–2009: U.S. Geological Survey Scientific Investigations Report 2011–5082, 84 p.

Contents

Contents—Continued

Contents—Continued

Figures

Figures—Continued

Tables

Conversion Factors and Datums

Multiply	By	To obtain
inch (in.)	2.54	centimeter (cm)
inch (in.)	25.4	millimeter (mm)
foot (ft)	0.3048	meter (m)
mile (mi)	1.609	kilometer (km)
square mile (mi^2)	2.590	square kilometer (km^2)

Horizontal coordinate information is referenced to the North American Datum of 1983 (NAD 83).

Concentrations of chemical constituents in water are given either in milligrams per liter (mg/L, approximately equivalent to parts per million) or micrograms per liter (µg/L, approximately equivalent to parts per billion).

Occurrence and Distribution of Pesticides in Surface Waters of the Hood River Basin, Oregon, 1999–2009

By Whitney B. Temple and Henry M. Johnson

Abstract

The U.S. Geological Survey analyzed pesticide and trace-element concentration data from the Hood River basin collected by the Oregon Department of Environmental Quality (ODEQ) from 1999 through 2009 to determine the distribution and concentrations of pesticides in the basin's surface waters. Instream concentrations were compared to (1) national and State water-quality standards established to protect aquatic organisms and (2) concentrations that cause sublethal or lethal effects in order to assess their potential to adversely affect the health of salmonids and their prey organisms. Three salmonid species native to the basin are listed as "threatened" under the U.S. Endangered Species Act: bull trout, steelhead, and Chinook salmon.

A subset of 16 sites was sampled every year by the ODEQ for pesticides, with sample collection targeted to months of peak pesticide use in orchards (March–June and September). Ten pesticides and four pesticide degradation products were analyzed from 1999 through 2008; 100 were analyzed in 2009. Nineteen pesticides were detected: 11 insecticides, 6 herbicides, and 2 fungicides. Two of four insecticide degradation products were detected. All five detected organophosphate insecticides and the one detected organochlorine insecticide were present at concentrations exceeding water-quality standards, sublethal effects thresholds, or acute toxicity values in one or more samples. The frequency of organophosphate detection in the basin decreased during the period of record; however, changes in sampling schedule and laboratory reporting limits hindered clear analysis of detection frequency trends. Detected herbicide and fungicide concentrations were less than water-quality standards, sublethal effects thresholds, or acute toxicity values. Simazine, the most frequently detected pesticide, was the only herbicide detected at concentrations within an order of magnitude (factor of 10) of concentrations that impact salmonid olfaction. Some detected pesticides are of concern, not for their toxicity alone, but for their ability to potentiate the harmful impacts of other pesticides, particularly organophosphates, on salmonids or their prey. Many samples contained mixtures of pesticides, but the effects to salmonids of relevant mixtures at environmentally realistic concentrations for the basin are unknown. Trace-element concentration data, although limited, indicate that eight trace elements are also of concern for their potential to harm salmonid health. The dataset is limited with regard to the spatial and seasonal distribution of pesticides and trace elements in all salmonid-bearing streams, the presence of particle-bound pesticides, and the presence of several unmonitored pesticides known to be used in the basin.

Introduction

Hood River drains 339 mi^2 on the northern side of Mt. Hood in Oregon and joins the Columbia River at the city of Hood River (fig. 1). Annual precipitation varies with topography, exceeding 110 in. in the southern, high elevation areas near Mt. Hood and averaging 30 in. on the valley floor near the city of Hood River. Most of the Hood River basin is forested and much of the remaining land is in agriculture (appendix A). Hood River is the largest city in the basin and has a population of 6,945 (U.S. Census Bureau, 2010). Agriculture, forest products, and tourism provide the economic base of the area.

Historically, the Hood River and its tributaries served as important spawning and rearing streams for anadromous and nonmigratory salmonids and for Pacific lamprey. Currently, three salmonids native to the Hood River basin are listed as "threatened" by the U.S. Fish and Wildlife Service (2010) under the U.S. Endangered Species Act—bull trout, steelhead, and Chinook salmon—in response to declining populations. The Pacific lamprey is a culturally significant fish for the native tribes along the Columbia River. As recently as 1963, Pacific lamprey were found throughout the basin (U.S. Department of Agriculture Forest Service, 1996). Their population has been limited to the lower 4.5 miles of Hood River since at least the mid-1990s. Three hundred seventy-three miles of streams in the Hood River basin are classified as critical habitat for salmonids (StreamNet, 2010). Instream passage barriers, flow modification, impaired water quality, and natural and anthropogenically induced sedimentation have been identified as contributors to the declining populations (Coccoli, 2004).

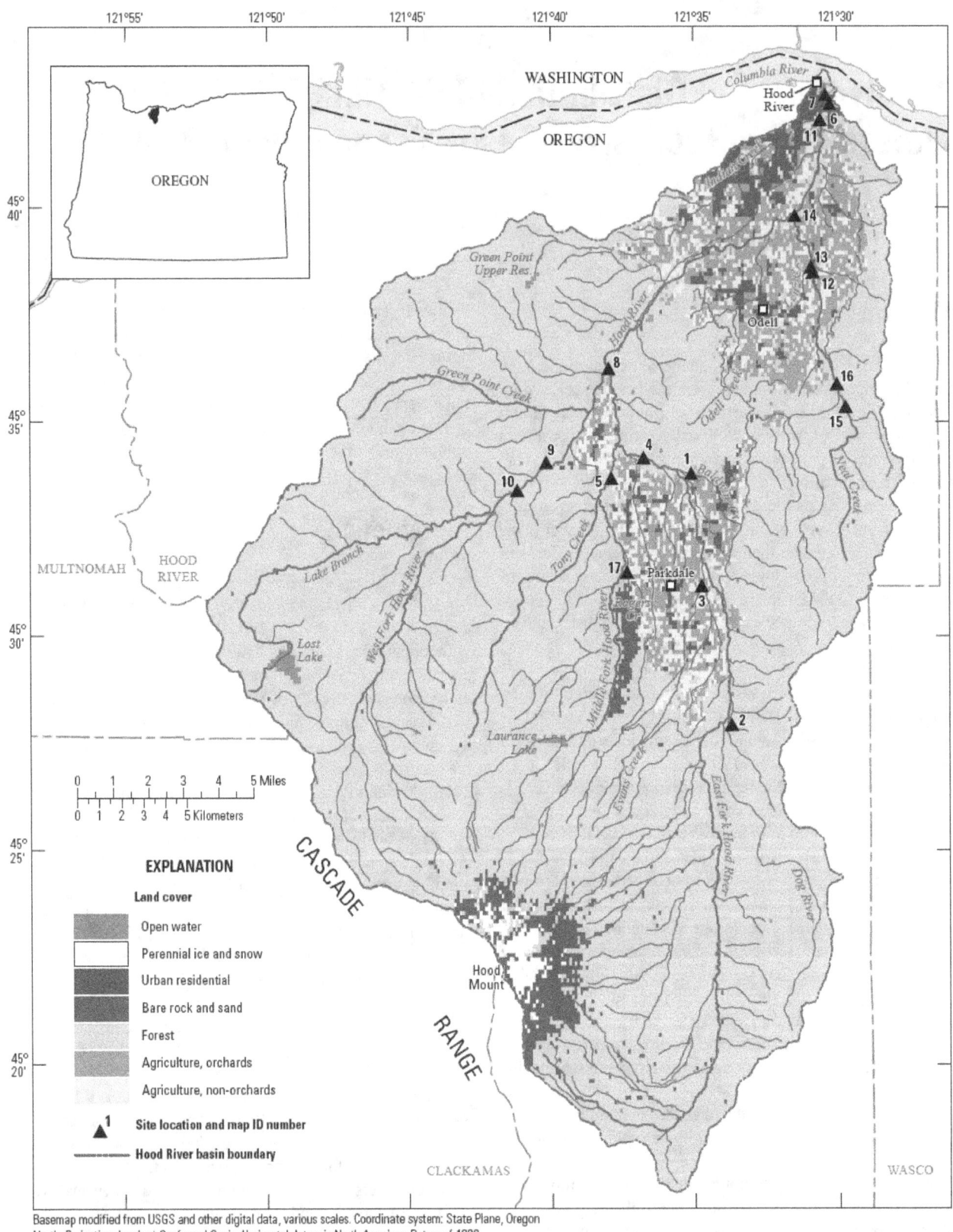

Basemap modified from USGS and other digital data, various scales. Coordinate system: State Plane, Oregon North, Projection: Lambert Conformal Conic. Horizontal datum is North American Datum of 1983.

Figure 1.   The Hood River basin, Oregon. Site names for the map ID numbers are in appendix A.

To begin to address impaired water quality related to agricultural activities, the Oregon Department of Environmental Quality (ODEQ) initiated a pesticide stewardship partnership (PSP) in the Hood River basin in 1999. Working in conjunction with growers, agricultural extension agents, the soil and water conservation district, watershed council, the Confederated Tribes of Warm Springs, and Oregon Department of Agriculture, the PSP seeks to "identify problems and improve water quality associated with pesticide use at the local level" (Oregon Department of Agriculture and others, 2008). Water was collected from streams throughout the Hood River basin and was analyzed for nine currently used pesticides. The organophosphate insecticides azinphos-methyl and chlorpyrifos were frequently detected in streams that flow through agricultural land at concentrations that periodically exceeded Oregon's acute or chronic water-quality standards (Coccoli, 2004). Monitoring has continued during the last decade as growers have implemented best management practices to try to minimize the offsite movement of pesticides and to reduce negative impacts to nontarget organisms.

Purpose and Scope

This report, prepared by the U.S. Geological Survey at the request of the Confederated Tribes of Warm Springs, summarizes the pesticide and trace element data collected by the ODEQ from 1999 through 2009 in the Hood River basin, Oregon, and uses the data to assess the potential effects of these contaminants on the health of salmonids that spawn and spend the first years of their life in the basin. Concentrations of pesticides and trace elements are compared to water-quality standards and mortality endpoints for fish and aquatic invertebrates, where such standards and endpoints exist. Concentrations of pesticides and trace elements also are compared to sublethal effects on salmonids documented in peer-reviewed literature and U.S. Environmental Protection Agency databases. Finally, the report identifies gaps in information about the occurrence of contaminants toxic to salmonids and aquatic invertebrates in streams of the Hood River basin.

Methods

Concentrations of pesticides and trace elements used for this report were obtained from the ODEQ's Laboratory Analytical Storage and Retrieval (LASAR) database (Oregon Department of Environmental Quality, 2008). Only primary (not quality assurance) samples were analyzed. The data were retrieved on August 9, 2010. The pesticide data were

organized into two datasets. The trace element data were organized into a third dataset. Each is briefly described below. Common chemical names for pesticide active ingredients are used in this report and differ in some cases from the names used in the ODEQ's database. The ODEQ's database includes some pesticide product names and alternate spellings of chemical names, Baygon for propoxur, Imidan for phosmet, Guthion for azinphos-methyl, oxygen analog for oxon, and chlorpyriphos or Dursban for chlorpyrifos, for example.

Pesticide Concentration Data, Ambient Stream Water, 1999–2009

Data were collected at 16 sites in the Hood River basin from 1999 through 2009. Catchment area and land use information for those sites are provided in appendix A. Sample counts for all sites are shown in table 1. Nine hundred fifty-three (953) surface-water pesticide samples were collected between 1999 and 2009. Most samples were collected during the spring and summer, to coincide with orchard pesticide application in the basin (78 percent of samples were collected March–June). Data from 1999 through 2002 came from a Portland State University and ODEQ study on the effects of instream exposure to pesticides on threatened steelhead in the Hood River basin (Eugene Foster, Portland State University, written commun., 2003; Oregon Department of Environmental Quality, 2008). That study examined nine pesticides at eight sites. The remaining water-quality data from routinely monitored sites came from the Hood River Pesticide Stewardship Partnership (HRPSP) project, which included as many as 10 organophosphate insecticides, 4 organophosphate degradation products, and 2 herbicides at a subset of the 16 sites through 2008. In 2009, an expanded list of 100 pesticides was analyzed for 8 of those sites to account for changes in pesticides used in the basin. Appendix B shows the number of samples collected at each site during each month and year.

Pesticide Concentration Data, Special Study on Effluent from Fruit Packers, 2004–2005

Fifty pesticide samples from 37 surface-water sites in the Hood River basin were collected in 2004 (n = 12) and 2005 (n = 38) for the National Pollutant Discharge Elimination System permit monitoring project for fruit packing facilities. Ten pesticides were analyzed in that study. Samples from 13 sites are from effluent from packing plants, and the remaining samples are from stream water. Sample counts by site are shown in table 1, with the receiving stream for effluent sites listed. Surface-water samples were collected by ODEQ staff at, upstream, and downstream of fruit-packing-plant discharge sites.

Table 1. Number of samples and period of record for sampling sites in the Hood River basin, Oregon, 1999–2009.

[**Abbreviations:** HRPSP, Hood River Pesticide Stewardship Partnership; ODEQ, Oregon Department of Environmental Quality; ODFW, Oregon Department of Fish and Wildlife; Ppl, Pacific Power and Light; PSU, Portland State University; RM, river mile; RR, railroad; VH, Van Horn; –, no samples or not applicable]

ODEQ station ID	Site name (full)	Site name (short)	Number of water samples from source				Data period	Total samples	Stream receiving fruit processing effluent
			HRPSP	PSU study	Fruit packers study	Trace elements			
13141	Neal Creek at mouth (upstream of bridge)	Neal, mouth	114	48	–	40	1999-2009	202	–
11972	Lenz Creek at mouth	Lenz	79	35	–	29	1999, 2001-06, 2008-09	143	–
25123	Upper Neal Creek above agriculture diversion	Neal, upper, above diversion	76	37	–	26	2001-07	139	–
25124	Evans Creek at bridge (Baseline Road)	Evans	64	19	–	24	2001-06	107	–
30174	Upper Neal Creek, downstream	Neal, upper, below diversion	97	–	–	–	2003-09	97	–
13158	Hood River downstream of Ppl Powerdale Powerhouse	Hood, mouth	63	7	–	13	1999-01	83	–
31499	Middle Neal Creek at Hwy 35	Neal, middle	73	–	–	–	2004-09	73	–
25133	Dog River below Puppy Creek confluence	Dog	23	18	–	22	2001-04	63	–
13138	East Fork Hood River at County Gravel Pit (River Mile 0.75)	Hood, East Fork	–	32	–	21	1999, 2000, 2002	53	–
13181	Baldwin Creek at end of Baldwin Creek Road	Baldwin	52	–	–	–	2003-06	52	–
13140	West Fork Hood River at Lost Lake Road (River Mile 4.7)	Hood, West Fork, RM 4.7	20	–	–	14	1999-01	34	–
13139	Middle Fork Hood River at River Mile 1.0 (ODFW Smolt Trap)	Hood, Middle Fork	17	–	–	14	1999-00	31	–
12012	Hood River at footbridge downstream of I-84	Hood, mouth	20	–	–	7	2002, 2005-09	27	–
34788	Rogers Spring Creek at Red Hill Driver (RM 0.25)	Rogers	27	–	–	–	2008-09	27	–
10681	West Fork Hood River at mouth	Hood, West Fork, mouth	21	–	–	–	2008-09	21	–
31505	Lenz Creek Packing Plant		–	–	7	–	2004	7	Lenz
34787	West Fork Hood River at Moving Falls (RM 2.5)	Hood, West Fork, RM 2.5	6	–	–	–	2008-09	6	–
21634	Indian Creek near mouth	Indian	–	5	–	–	1999-00	5	–
11968	McGuire Creek upstream of Diamond-Odell Plant (discharge #2)	–	–	–	2	1	1999, 2005	3	Odell
11971	Neal Creek upstream of Lenz Creek confluence	–	–	–	2	1	1999, 2005	3	Neal
11973	Stadelman-Whitney discharge at end of ditch	–	–	–	2	1	1999, 2004-05	3	Lenz
13173	McGuire Creek at John Weber Park	–	–	–	2	1	1999, 2005	3	Odell
22966	Lage Orchard Packing Plant outfall	–	–	–	2	1	1999, 2004-05	3	unnamed Whiskey Creek tributary

Table 1. Number of samples and period of record for sampling sites in the Hood River basin, Oregon, 1999–2009.—Continued

[**Abbreviations:** HRPSP, Hood River Pesticide Stewardship Partnership; ODEQ, Oregon Department of Environmental Quality; ODFW, Oregon Department of Fish and Wildlife; Ppl, Pacific Power and Light; PSU, Portland State University; RM, river mile; RR, railroad; VH, Van Horn; –, no samples or not applicable]

| ODEQ station ID | Site name (full) | Site name (short) | Number of water samples from source | | | | Data period | Total samples | Stream receiving fruit processing effluent |
			HRPSP	PSU study	Fruit packers study	Trace elements			
11967	Diamond Fruit-Odell Plant effluent (discharge #2)	–	–	–	1	1	1999, 2004	2	Odell
11969	McGuire Creek downstream of Diamond-Odell discharge #2 & downstream of RR tracks	–	–	–	1	1	1999, 2005	2	Odell
11974	Duckwall final effluent	–	–	–	1	1	1999, 2005	2	Lenz
13171	Lenz Creek at Stadleman Drive Pump Station	–	–	–	1	1	1999, 2005	2	Lenz
13172	McGuire Creek at Davis Drive	–	–	–	1	1	1999, 2005	2	Odell
22965	Lage Orchard Plant receiving ditch upstream of packing plant	–	–	–	1	1	1999, 2005	2	unnamed Whiskey Creek tributary
22970	Odell Creek at Ehrck Hill Road	–	–	–	1	1	1999, 2005	2	Odell
22971	Diamond-Central Plant, discharge at end of pipe	–	–	–	1	1	1999, 2005	2	Lenz
22972	Lenz Creek upstream of Duckwall-Pooley Odell Plant receiving	–	–	–	1	1	1999, 2005	2	Lenz
22973	Lenz Creek downstream of Duckwall-Pooley Plant receiving	–	–	–	1	1	1999, 2005	2	Lenz
22974	Lenz Creek downstream of Stadleman-Whitney Plant outfall	–	–	–	1	1	1999, 2005	2	Lenz
22978	Lenz Creek at Ehrck Hill Road	–	–	–	1	1	1999, 2005	2	Lenz
23940	Neal Creek downstream of Lenz Creek at RR bridge	–	–	–	–	2	3998	2	Neal
28831	Duckwall-Pooley Discharge downstream of RR tracks	–	–	–	2	–	2004-05	2	Lenz
32668	McGuire Creek downstream of Diamond-Odell discharge #2 ditch	–	–	–	2	–	2005	2	Odell
45918	Duckwall Pooley VH Plant receiving ditch at Hwy 3	–	–	–	–	2	3998	2	Neal
11410	Neal Creek at County Road	–	–	–	–	1	1999	1	Neal
11975	Ditch upstream of Diamond-Central (Lenz Creek tributary)	–	–	–	–	1	1999	1	Lenz
11976	Ditch downstream of Diamond-Central (Lenz Creek tributary)	–	–	–	–	1	1999	1	Lenz
13167	Wisehart Creek at Woodworth Drive	–	–	–	–	1	1999	1	Wisehart
13174	Odell Creek at John Weber Park upstream of McGuire Creek	–	–	–	–	1	1999	1	Odell
22956	Lage Orchard Packing Plant receiving stream downstream at Bend N	–	–	–	–	1	1999	1	Neal

Table 1.   Number of samples and period of record for sampling sites in the Hood River basin, Oregon, 1999–2009.—Continued

[**Abbreviations:** HRPSP, Hood River Pesticide Stewardship Partnership; ODEQ, Oregon Department of Environmental Quality; ODFW, Oregon Department of Fish and Wildlife; Ppl, Pacific Power and Light; PSU, Portland State University; RM, river mile; RR, railroad; VH, Van Horn; –, no samples or not applicable]

| ODEQ station ID | Site name (full) | Site name (short) | Number of water samples from source | | | | Data period | Total samples | Stream receiving fruit processing effluent |
			HRPSP	PSU study	Fruit packers study	Trace elements			
22957	Lage Orchard Plant receiving stream at Hwy 35	–	–	–	–	1	1999	1	Neal
22958	Lage Orchard Plant receiving stream tributary upstream of main receiving stream	–	–	–	–	1	1999	1	Neal
22960	Diamond-Parkdale Plant, outfall at plant	–	–	–	–	1	1999	1	Wisehart
22961	Diamond-Parkdale Plant receiving ditch 50 feet downstream	–	–	–	–	1	1999	1	Wisehart
22962	Diamond-Parkdale receiving ditch upstream of Wisehart Creek	–	–	–	–	1	1999	1	Wisehart
22963	Wisehart Creek upstream of Diamond-Parkdale receiving ditch	–	–	–	–	1	1999	1	Wisehart
22964	Wisehart Creek 50 feet downstream of Diamond-Parkdale Plant receiving ditch	–	–	–	–	1	1999	1	Wisehart
22967	Lage Orchard Packing Plant receiving stream 50 feet downstream of plant	–	–	–	–	1	1999	1	Neal
22968	Diamond-Odell Plant discharge #1 at end of Pipe M	–	–	–	–	1	1999	1	Odell
22969	Odell Creek upstream of Ehrck Hill Road behind church off Aga Road	–	–	–	–	1	1999	1	Odell
22975	Stadleman-Lenz Plant discharge at end of pipe	–	–	–	–	1	1999	1	Lenz
22976	Lenz Creek downstream of Stadleman-Lentz Plant outfall	–	–	–	–	1	1999	1	Lenz
22977	Lenz Creek downstream of Hanel Lumber Company	–	–	–	–	1	1999	1	Lenz
23002	Duckwall-Pooley Plant outfall at Van Horn	–	–	–	–	1	1999	1	Neal
23003	Duckwall-Pooley VH Plant receiving stream 300 yards downstream	–	–	–	–	1	1999	1	Neal
28828	Diamond-Central Discharge Pit	–	–	–	1	–	2004	1	Lenz
28829	Tributary to Lenz Creek from south downstream of Lingren Road	–	–	–	1	–	2005	1	Lenz
28830	Lenz Creek upstream of tributary from south (downstream of Lingren Rd)	–	–	–	1	–	2005	1	Lenz
28832	Duckwall-Pooley upstream of discharge at RR tracks	–	–	–	1	–	2005	1	Lenz

Table 1. Number of samples and period of record for sampling sites in the Hood River basin, Oregon, 1999–2009.—Continued

[**Abbreviations:** HRPSP, Hood River Pesticide Stewardship Partnership; ODEQ, Oregon Department of Environmental Quality; ODFW, Oregon Department of Fish and Wildlife; Ppl, Pacific Power and Light; PSU, Portland State University; RM, river mile; RR, railroad; VH, Van Horn; –, no samples or not applicable]

ODEQ station ID	Site name (full)	Site name (short)	Number of water samples from source				Data period	Total samples	Stream receiving fruit processing effluent
			HRPSP	PSU study	Fruit packers study	Trace elements			
28834	Unnamed Pine Grove tributary downstream of Hwy 35, upstream of other road ditch	–	–	–	1	–	2005	1	Neal
28851	Lenz Creek upstream of Stadleman-Whitney discharge ditch	–	–	–	1	–	2005	1	Lenz
28852	Stadelman-Whitney discharge at plant	–	–	–	1	–	2005	1	Lenz
28853	Lenz Creek upstream of Stadelman-Whitney discharge	–	–	–	1	–	2005	1	Lenz
28856	Pine Grove tributary upstream of Neal Creek	–	–	–	1	–	2005	1	Neal
28857	Neal Creek upstream of Pine Grove tributary	–	–	–	1	–	2005	1	Odell
32572	Hood River upstream of Neil Creek	–	–	–	1	–	2005	1	Hood
32573	Hood River downstream of Neil Creek, right bank	–	–	–	1	–	2005	1	Hood
32663	Moore Orchard effluent	–	–	–	1	–	2005	1	Neal
32664	Wells Orchard effluent	–	–	–	1	–	2005	1	unnamed Whiskey Creek tributary
32676	McGuire Creek at Odell Hwy	–	–	–	1	–	2005	1	Odell
32775	Diamond-Central effluent discharge in manhole at plant	–	–	–	1	–	2005	1	Lenz
32776	Diamond-Odell Plant effluent discharge #2 at end of pipe	–	–	–	1	–	2005	1	Odell
34370	Hood River Juice Company, holding pond	–	–	–	–	1	2009	1	Indian
36059	Hood River Juice Co Surface Water in ditch under drainpipe area (down slope from road)	–	–	–	–	1	2009	1	Indian
36060	Hood River Juice Co Farmer's Irrigation Ditch Downslope of road and ditch	–	–	–	–	1	2009	1	Indian

Trace Element Concentration Data, 1999–2009

Two hundred fifty-five (255) surface-water samples collected from 53 sites in the Hood River basin during 1999–2002 and 2009 were analyzed for trace element concentrations. These data came from the ODEQ's LASAR database, although they were not all collected as part of the HRPSP project. Twenty-seven trace elements were analyzed. Most sites were sampled once; some had as many as 40 samples.

Reporting Limits and Data Screening

The reporting limit (RL) is the value at which a laboratory reports a concentration as undetected. The value is said to be censored at that RL. No determination can be made about the magnitude of a concentration less than the RL; it might be slightly less than the RL or it might be zero. The RL is based on the laboratory's analysis of several types of quality control samples, including blanks, replicates, matrix spikes, and surrogate spikes. A laboratory's ability to quantify a concentration often changes over time as a result of changing laboratory techniques, equipment, and analysts, and also because of sample-to-sample differences in water chemistry. As a result, RLs change over time.

From 1999 through 2009, RLs for all pesticides varied among and within years. Figures 2 and 3 show reporting limits by year for azinphos-methyl and simazine, respectively, to

exemplify the variability in reporting limits in this dataset. Comparing samples with different reporting limits can misrepresent the frequency and distribution of occurrence. Pesticides may seem to be more widely distributed or to be detected more frequently during periods of time with lower RLs compared to periods with higher RLs.

To address the issue of multiple RLs, the pesticide data from 1999 through 2009 were screened for some analyses in this report. Often, the screening level is set equal to the highest RL in a dataset. This technique would have resulted in the loss of a large amount of information in this dataset. Instead, the screening level was set to minimize the loss of data. The screening level for each pesticide is shown in table 2. Positive detections less than the screening level were censored at the screening level. Censored values less than the screening level were recensored at the screening level. Data censored at a RL greater than the screening level were removed from the dataset. Similarly, positive detections collected during periods when the RL was greater than the screening level also were removed to avoid biasing the data. Four hundred four (404) data points were removed from the screened dataset. Of those, only 2 were detected concentrations.

Screened and unscreened data were used in different sections of this report. Each section or analysis includes a description indicating which data were used.

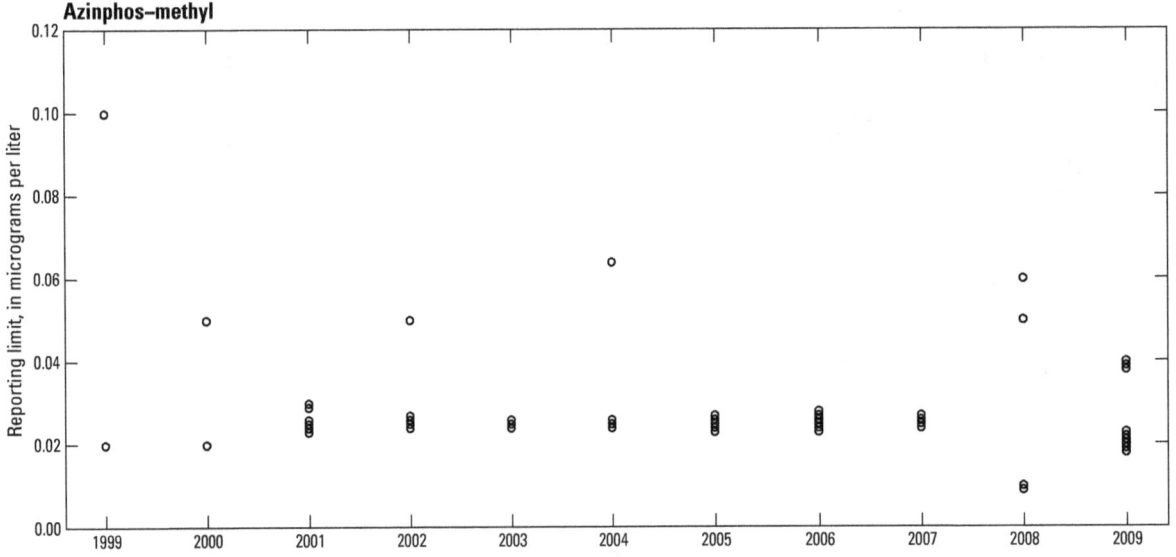

Figure 2.   Reporting limits by year for azinphos-methyl in samples from the Hood River basin, Oregon.

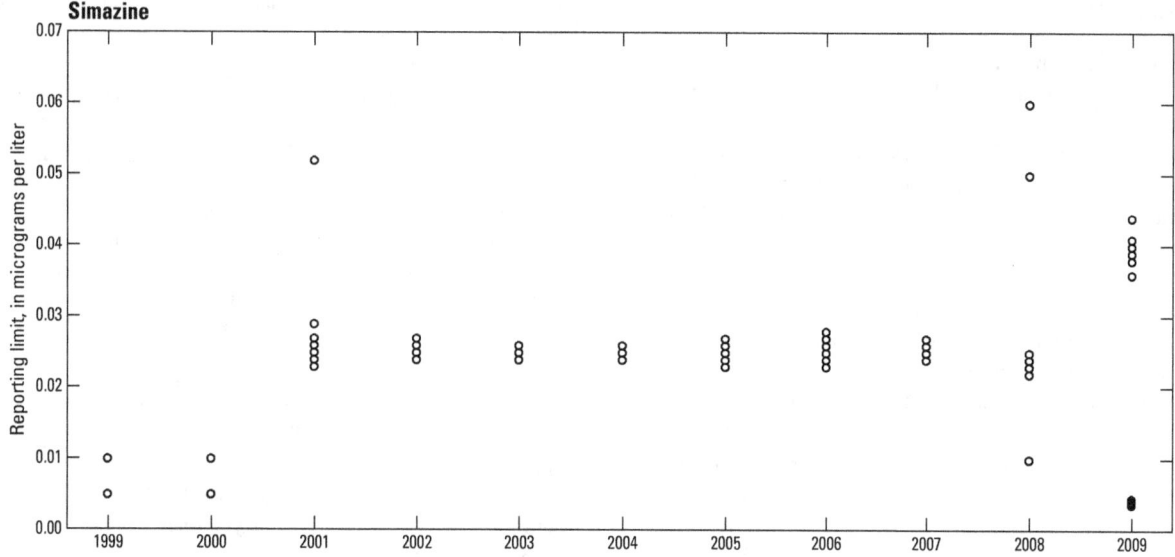

Figure 3.   Reporting limits by year for simazine in samples from the Hood River basin, Oregon.

Table 2.   Screening level and reporting limit ranges by pesticide.

[**Abbreviations:** μg/L, micrograms per liter]

Pesticide	Screened reporting limit (μg/L)	Reporting limit range (μg/L)
Atrazine	0.027	0.0036–0.11
Azinphos-methyl	0.03	0.009–0.1
Carbaryl	0.005	0.0045–0.0057
Chlorpyrifos	0.03	0.009–0.1
DEET	0.0051	0.0045–0.0057
Diazinon	0.1	0.009–0.113
Diuron	0.004	0.0037–0.0045
Endrin	0.063	0.054–0.068
Fluometuron	0.0043	0.0037–0.0045
Hexazinone	0.04	0.0036–0.0045
Imidacloprid	0.023	0.018–0.023
Malathion	0.03	0.009–0.1
Methomyl	0.0021	0.0018–0.0023
Norflurazon	0.023	0.018–0.023
Phosmet	0.05	0.009–0.064
Propiconazole	0.023	0.018–0.023
Propoxur	0.0021	0.0018–0.0023
Pyraclostrobin	0.0045	0.0038–0.0045
Simazine	0.027	0.0037–0.064

Pesticide Use

Watershed-level pesticide-use information is difficult to obtain. The mid-Columbia basin, which includes the Hood River basin, had the highest reported amount (by weight) of pesticides applied across Oregon in 2007 and 2008 (55 and 38 percent of the total amount reported statewide, respectively) (Oregon Department of Agriculture, 2008, 2009). The high amount of pesticides used in the mid-Columbia basin was driven by the heavy use of metam sodium, a soil fumigant (Oregon Department of Agriculture, 2008, 2009). Soil fumigants often account for a large proportion of pesticide use when reported by weight due to their high application rates relative to other types of pesticides and are more heavily used on crops grown elsewhere in the mid-Columbia basin than on orchards (U.S. Environmental Protection Agency, 2005d). Total acreage treated with chemicals to control insects, weeds, grass, brush, or diseases in crops and orchards, and to control growth, thin fruit, ripen, or defoliate crops in Hood River County increased from 2002 to 2007 (U.S. Department of Agriculture, 2007). Information on pesticides registered and deemed appropriate for use by crop are available from Oregon State University Extension Service's 2009 Integrated Pest Management Handbooks for Weeds, Insects, and Diseases (Hollingsworth, 2009; Peachey, 2009; Pscheidt and Ocamb, 2009) and 2010 Pest Management Guide for Tree Fruits in the mid-Columbia Area (Oregon State University Extension Service, 2010). Appendix C lists pesticides considered suitable for the major land uses of the Hood River basin.

Agriculture

Table 3 lists pesticides known to be commonly used on agricultural crops in the Hood River basin at the time of the writing of this report (Steve Castagnoli, Oregon State University Extension Service, oral commun., 2010). The list includes herbicides for weed control, insecticides (including miticides and acaracides) for arthropod control, and fungicides to control various blights, rusts, and molds. Numerous others are registered for use on crops grown in the Hood River basin (appendix C). Most of the pesticides in table 3 and appendix C were not analyzed for this project, which focused on organophosphates through 2008 due to the existence of State water-quality standards for those pesticides. Agricultural pesticide use in the basin varies across years in response to changes in pest occurrence and to reduce the potential for pesticide resistance.

Forestry

Herbicides are the most common class of pesticides used in forests. Overall, herbicide use fluctuates across sites and years in order to meet localized needs (Doug Thiesies, Oregon Department of Forestry, oral commun., 2010). Forestry herbicides commonly used in the Hood River basin are listed in table 4. Sulfometuron methyl, glyphosate, and 2,4-D are commonly used in the fall for site preparation (Doug Thiesies, Oregon Department of Forestry, oral commun., 2010); however, only 2,4-D was analyzed in this project. In the Pacific Northwest, 2,4-D, glyphosate, imazapyr, picloram, or triclopyr are used for nearly all brush and weed tree control, although other pesticides are registered for this purpose (Peachey, 2009). Hexazinone is used in the spring at the time of planting (Doug Thiesies, Oregon Department of Forestry, oral commun., 2011) and was analyzed in this project in 2009. Insecticide use is rare, although widespread applications may occur in response to the outbreak of a specific pest (Doug Thiesies, Oregon Department of Forestry, oral commun., 2010).

Rights-of-Way

Bromacil, 2,4-D ester, diuron, glyphosate, sulfometuron methyl, and triclopyr are known to have been used for weed and brush control adjacent to roads or irrigation canals since 2009 (Brian Walker, Oregon Department of Transportation, oral commun., 2010; John Buckley, East Fork Irrigation District, oral commun., 2010; Nate Lain, Hood River County Weed and Pest Division, oral commun., 2010). 2,4-D, bromacil, diuron, and triclopyr were analyzed for this project in 2009.

Table 3. Pesticides commonly used on agricultural crops in the Hood River basin, 2009–10.

Chemical class	Pesticide
Fungicides	
Azole	Myclobutanil
Azole	Triflumizole
Benzimidazole precursor	Thiophanate-methyl
Beta-methoxyacryl ester	Beta-methoxyacryl ester
Dithiocarbamate	Mancozeb
Dithiocarbamate	Ziram
Inorganic copper	Copper products
Strobilurin	Pyraclostrobin
Herbicides	
2,6-Dinitroaniline	Oryzalin
Amide	Pronamide
Diphenyl ether	Oxyfluorfen
Phosphonoglycine	Glyphosate
Substituted urea	Diuron
Triazine	Simazine
Insecticide	
Anthranilic diamide	Rynaxypyr
Avermectin	Abamectin
Inorganic sulfur	Sulfur products
Unclassified	Pyriproxyfen
Neonicotinoid	Acetamiprid
Neonicotinoid	Clothianidin
Neonicotinoid	Imidacloprid
Neonicotinoid	Thiacloprid
Pyrethroid	Deltamethrin
Pyrethroid	Fenpropathrin
Pyrethroid	Gamma-cyhalothrin
Pyrethroid	Lambda-cyhalothrin
Pyrethroid	Permethrin
Spinosyn	Spinetoram

Table 4. Herbicides commonly used for forest management in the Hood River basin, 2010.

Herbicides		
2,4-D	Hexazinone	Metsulfuron methyl
Clopyralid	Imazapyr	Sulfometuron methyl
Glyphosate	Metsulfuron	Triclopyr ester

Household Use

Household use of pesticides is difficult to assess. The Oregon Department of Agriculture's Pesticide Use Reporting System (PURS) includes an annual survey of about 1,500 Oregon households regarding pesticide use. However, PURS has been suspended since 2009 due to budget constraints. Combined data for Hood River and Wasco counties indicate that herbicides were the primary class of pesticides used in households in 2007 and 2008 (99.4 and 81.3 percent, respectively) (Oregon Department of Agriculture, 2008, 2009). Statewide, glyphosate and 2,4-D accounted for approximately 78 percent of reported household herbicide use in 2007 and 2008. Fipronil and S-methoprene (2007) and malathion (2008) were the insecticides with the highest reported use statewide. Both years, DEET (N,N-Diethyl-meta-toluamide) was by far the most common form of insect repellent used. Calcium polysulfide and Captan accounted for 62 to 74 percent of reported fungicide use in households.

Results

Pesticide Concentration Data from Ambient Stream Water, 1999–2009

As part of the following discussions of each detected pesticide, the range of detected concentrations is compared to established water-quality standards and mortality and sublethal effect values. Sublethal effects are physiological or behavioral changes that occur to an organism after exposure to a contaminant at a less-than-fatal concentration. In figures 4–22, each sublethal effect value from the literature is represented by a single black square and each no observed effect concentration (NOEC) value is shown by an "X". Multiple values are shown for a given sublethal endpoint when multiple values were available from the literature and may reflect differences in test species, product formulation, exposure duration, or other variations in experimental design. The established water-quality standards are freshwater values from the ODEQ 2004 water-quality criteria, U.S. Environmental Protection Agency (USEPA) national recommended water-quality criteria, and USEPA Office of Pesticide Programs' aquatic life benchmarks (appendix D) (Oregon Department of Environmental Quality, 2004; U.S. Environmental Protection Agency, 2005a, 2009b). Toxicity and sublethal endpoint data were obtained from USEPA Reregistration Eligibility Decisions (U.S. Environmental Protection Agency, 1994,

1996, 1997, 1998a-b, 1999, 2003, 2005e, 2006a-g, 2009d), USEPA pesticide fact sheets and ecological risk assessments (U.S. Environmental Protection Agency, 2001, 2005c, 2008, 2009a), National Marine Fisheries Service biological opinions (National Marine Fisheries Service, 2008, 2009), USEPA ECOTOX database (U.S. Environmental Protection Agency, 2007), U.S. Office of Pesticide Programs Pesticide Toxicity database (U.S. Environmental Protection Agency, 2005b), U.S. Geological Survey Columbia Environmental Research Center Acute Toxicity database (U.S. Geological Survey, 2004), pesticide product labels (Gowan Company, 2004; BASF Corporation, 2010), a National Pesticide Information Center fact sheet (National Pesticide Information Center, 2010), and selected peer-reviewed literature (Julin and Sanders, 1977; Spehar and others, 1981; Mayer and Ellersieck, 1988; Sheedy and others, 1991; Beketov and Liess, 2008; Tierney and others, 2010). Fish toxicity and sublethal effect data are for salmonids (*Oncorhynchus spp.* and *Salmo spp.*), except for a few specified values for fathead minnows (*Pimephales promelas*), which were used where no salmonid data exist. Fish toxicity values are for 96-hour tests. Invertebrate endpoints were selected for common toxicity test species that are likely to occur in the Pacific Northwest; for example, various species of stonefly, mosquito, scud, and zooplankton. Invertebrate toxicity test procedures are more variable than those for fish. Exposure durations for invertebrate toxicity tests were generally 24, 48, or 96 hours. This accounts for some of the variability in the salmonid prey toxicity values provided for a given pesticide.

A summary of the use and environmental fate of all detected pesticides is provided in appendix E. A complete list of all pesticides analyzed is provided in appendix F.

Pesticide Occurrence for Pesticides Analyzed 1999–2009

From 1999 through 2009, two herbicides and five insecticides were detected in the basin (table 5). At least one pesticide was detected at 13 of 16 sampling sites. The only sites without a pesticide detection were West Fork Hood River at Moving Falls (RM 2.5), West Fork Hood River at Lost Lake Road (River Mile 4.7), and Dog River below Puppy Creek confluence.

Concentrations at eight sites exceeded the USEPA or Oregon water-quality standards (table 6) for one or more pesticides at least once during the project period. However, because the standards are based on specific exposure durations (acute=24 hours, chronic=96 hours), point-in-time samples are not directly comparable to the standards.

Table 5.   Counts of samples and detections for pesticides analyzed from 1999 to 2009, sites where pesticides were detected, and concentration, date, and site of the maximum detected concentration for the entire period of record and for 2009.

[Sample size includes all (unscreened) samples. µg/L, micrograms per liter; –, not detected in 2009]

Pesticide type	Class	Pesticide	Number of samples	Number of detec- tions	Sites where detected	Maximum concentration 1999–2009			Maximum concentration 2009		
						Measure- ment (µg/L)	Date	Site	Measure- ment (µg/L)	Date	Site
Herbicide	Triazine	Atrazine	920	2	Lenz	0.032	05-31-02	Lenz	–	–	–
		Simazine	933	157	Baldwin; Evans; Hood, East Fork; Hood, mouth; Hood, Indian; Lenz; Neal, middle; Neal, mouth; Neal, upper, below diversion; Rogers	1.9	06-13-03	Lenz	0.299	09-16-09	Lenz
Insecticide	Organo- phosphate	Azinphos methyl	939	76	Evans; Hood, Middle Fork; Hood, mouth; Lenz; Neal, middle; Neal, mouth; Neal, upper, above diversion	0.375	09-12-03	Lenz	0.028	09-16-09	Lenz
		Chlorpyri- fos	944	66	Evans; Hood, East Fork; Hood, Middle Fork; Hood, mouth; Indian; Lenz; Neal, middle; Neal, mouth; Neal, upper, above diversion; Rogers	0.482	03-22-99	Neal, mouth	–	–	–
		Diazinon	945	4	Baldwin; Neal, mouth	0.19	06-20-00	Neal, mouth	–	–	–
		Malathion	946	7	Lenz; Neal, mouth	0.098	06-16-04	Lenz	–	–	–
		Phosmet	895	14	Evans; Lenz; Neal, middle; Neal, mouth; Neal, upper, above diversion	0.278	06-13-02	Lenz	–	–	–

Table 6. Percentage of samples exceeding U.S. Environmental Protection Agency or Oregon freshwater aquatic life standards, 1999–2009.

[Sample size includes screened samples, not total samples. All concentrations reported in micrograms per liter; **Abbreviations:** USEPA, U.S. Environmental Protection Agency; CMC, criteria maximum concentration; CCC, criterion continuous concentration; %, percent; —, no water-quality standard; <, less than]

| Pesticide | USEPA Office of Pesticide Programs Aquatic Life Benchmarks | | | | USEPA Water Quality Criteria | | Oregon Water Quality Criteria | | Percentage exceeding lowest standard (%) |
| | Fish | | Invertebrates | | | | | | |
	Acute	Chronic	Acute	Chronic	CMC (Acute)	CCC (Chronic)	Acute	Chronic	
Atrazine	2,650	65	360	60	–	–	–	–	0
Azinphos-methyl[1]	0.18	0.055	0.08	0.036	–	–	–	0.01	7
Sites where detected Evans									2
Lenz									26
Neal, middle									3
Neal, mouth									19
Neal, upper, above diversion									0.9
Chlorpyrifos	0.9	0.57	0.05	0.04	0.083	0.041	0.083	0.041	3
Sites where detected Evans									2
Hood, mouth									1
Indian									100
Lenz									6
Neal, middle									2
Neal, mouth									11
Diazinon	45	< 0.55	0.105	0.17	0.17	0.17	–	–	0.3
Sites where detected Baldwin									4
Neal, mouth									1
Malathion	0.295	0.014	0.005	0.000026	–	0.1	–	0.1	0.4
Sites where detected Lenz									0.9
Neal, mouth									2
Phosmet	35	3.2	1	0.8	–	–	–	–	0
Simazine	3,200	960	500	2,000	–	–	–	–	0

[1]Listed in Oregon Department of Environmental Quality (2004), Table 20 as Guthion (product name)

Occurrence, Detection Frequency, and Potential Impacts on Aquatic Life for Pesticides Analyzed 1999–2009

Additional information on the uses and environmental fate of each pesticide is provided in appendix C and appendix E.

Atrazine (herbicide)

Atrazine was detected in March 2003 and May 2004 at Lenz Creek at mouth. The detections were more than an order of magnitude (10 times) lower than concentrations known to cause changes in olfactory-mediated behavior and over three orders of magnitude (1,000 times) lower than the lowest aquatic life benchmark (fig. 4).

Azinphos-methyl (insecticide)

Azinphos-methyl was the second most frequently detected pesticide since sampling began in 1999 (n = 76 detections). Most detections occurred at Neal Creek at mouth (47 percent of detections) and Lenz Creek at mouth (43 percent). Most detections occurred in the summer and fall; more than 20 percent of samples collected during August through October had azinphos-methyl detected. During August to October, 53 percent of azinphos-methyl detections were in Lenz Creek; 47 percent were in Neal Creek (Middle or at mouth). Only six detections occurred in March–May, even though more than half of all samples were collected during those months. Twenty-one samples had detectable concentrations of azinphos-methyl oxon, a degradation product of azinphos-methyl. In 18 of those samples, azinphos-methyl and its degradation product were both detected. Detections of the degradation product were generally more common with higher measured concentrations of azinphos-methyl and were most common at Lenz Creek at mouth and Neal Creek at mouth. All detected concentrations of azinphos-methyl exceeded the Oregon chronic criterion (0.01 µg/L) and 37 percent of the detected concentrations exceeded the USEPA benchmark for acute exposures for invertebrates (0.08 µg/L). The highest detected concentrations were less than one order of magnitude lower than concentrations associated with sublethal changes to salmonids and mortality to their prey (fig. 5).

Chlorpyrifos (insecticide)

Chlorpyrifos was detected 66 times since 1999, last in 2008. The reporting limit in 2009 was higher than concentrations detected in 2008, which may account for the absence of detections in 2009. However, in 2009, chlorpyrifos was not detected at a concentration equal to or greater than 0.041 µg/L, the concentration of the lowest water-quality criterion, in 105 samples with reporting limits of 0.041 µg/L or less. The majority of detections were at Neal Creek at

mouth (47 percent) and Lenz Creek at mouth (18 percent). Forty-seven (47) of 66 detections occurred in March; the rest were in April. Chlorpyrifos-oxon, a degradation product of chlorpyrifos, was not detected in the 503 samples in which it was analyzed through 2008. Detections of chlorpyrifos were at concentrations exceeding the most stringent USEPA aquatic life benchmarks and at concentrations that can cause harmful effects to salmonids and their prey (fig. 6).

Diazinon (insecticide)

Diazinon was detected once at Neal Creek at mouth in June 2000 and March 2003 and twice in Baldwin Creek in June 2005. The detected concentrations exceeded the lowest USEPA benchmarks for acute and chronic exposures (for invertebrates) and were within a range known to cause invertebrate mortality and changes in salmonid olfaction (fig. 7).

Malathion (insecticide)

Malathion was detected seven times from 2000 through 2004 at concentrations that exceeded the minimum USEPA benchmarks for invertebrates and were within a range known to cause sublethal effects to invertebrates (fig. 8). Detections occurred in June and early July at Lenz Creek at mouth (n = 1) and Neal Creek at mouth (n = 6). Malathion-oxon, a degradation product, was detected in 3 of 460 samples through 2008, twice at Neal Creek at mouth and once at Lenz Creek at mouth. Of those, malathion was detected only in the sample with the highest malathion-oxon concentration.

Phosmet (insecticide)

Phosmet was detected in 14 samples from 2001 through 2008, mostly in Lenz Creek. Most detections (57 percent) occurred in the fall, even though most samples were collected during the spring and summer. The remaining 43 percent of detections were in April–June. Phosmet-oxon, a phosmet degradation product, was not detected in 459 samples in which it was analyzed. Phosmet detections were generally one order of magnitude lower than the minimum USEPA benchmarks, but overlapped the lower range of invertebrate toxicity values (fig. 9).

Simazine (herbicide)

With 157 detections from 1999 through 2009, simazine was the most commonly detected pesticide. Most detections occurred at Lenz Creek at mouth (44 percent of detections) and Neal Creek at mouth (43 percent). The highest concentrations and detection counts were in June, when 43 percent of simazine detections occurred. The maximum concentration was slightly less than a concentration known to cause olfactory changes in Atlantic salmon (fig. 10).

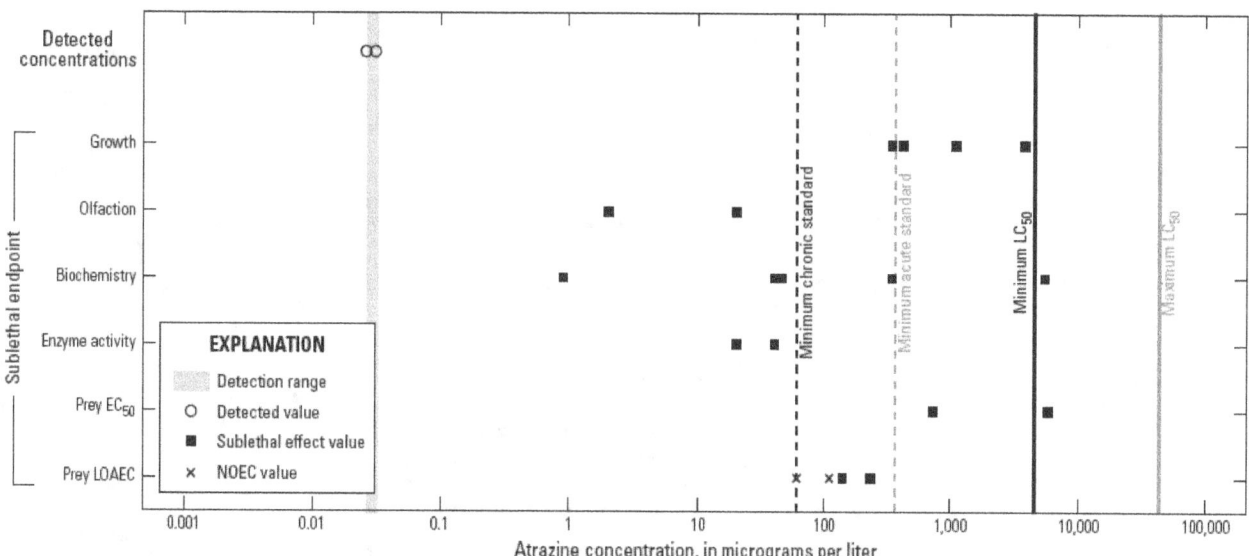

Figure 4. Detected atrazine concentrations in the Hood River basin, Oregon, compared to USEPA and Oregon water-quality standards and toxicity and sublethal endpoints for salmonids and their prey. Acute, 24-hour exposure; chronic, 96-hour exposure; LC$_{50}$, 50 percent lethal concentration; NOEC, no observed effect concentration; EC$_{50}$, 50 percent effective concentration; LOAEC, lowest observed adverse effect concentration.

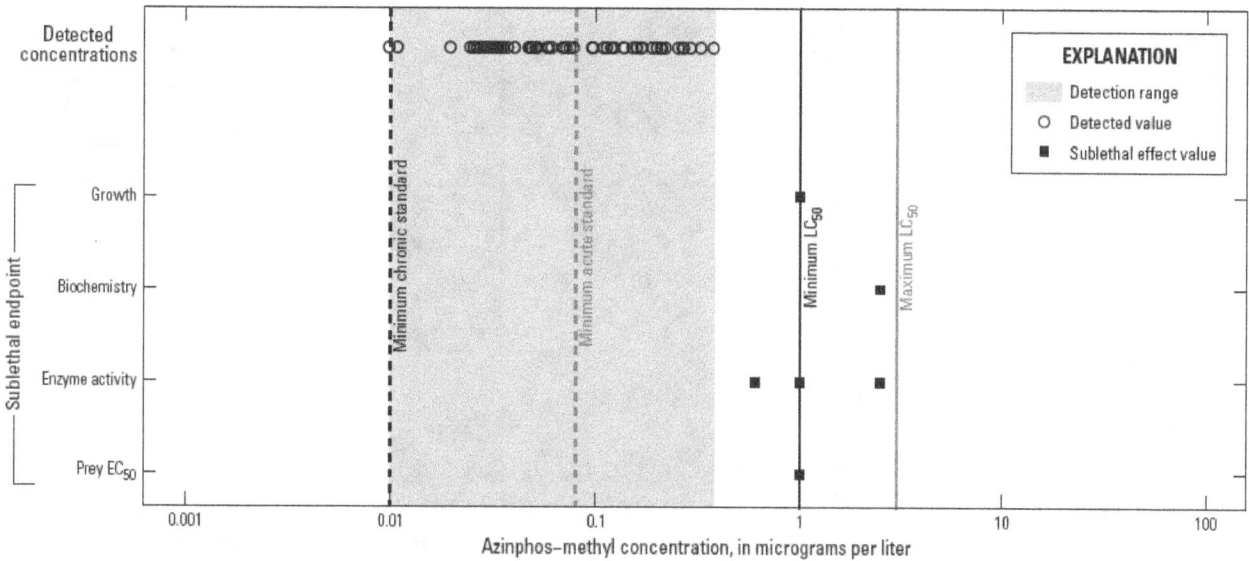

Figure 5. Detected azinphos-methyl concentrations in the Hood River basin, Oregon, compared to USEPA and Oregon water-quality standards and toxicity and sublethal endpoints for salmonids and their prey. Acute, 24-hour exposure; chronic, 96-hour exposure; LC$_{50}$, 50 percent lethal concentration; EC$_{50}$, 50 percent effective concentration.

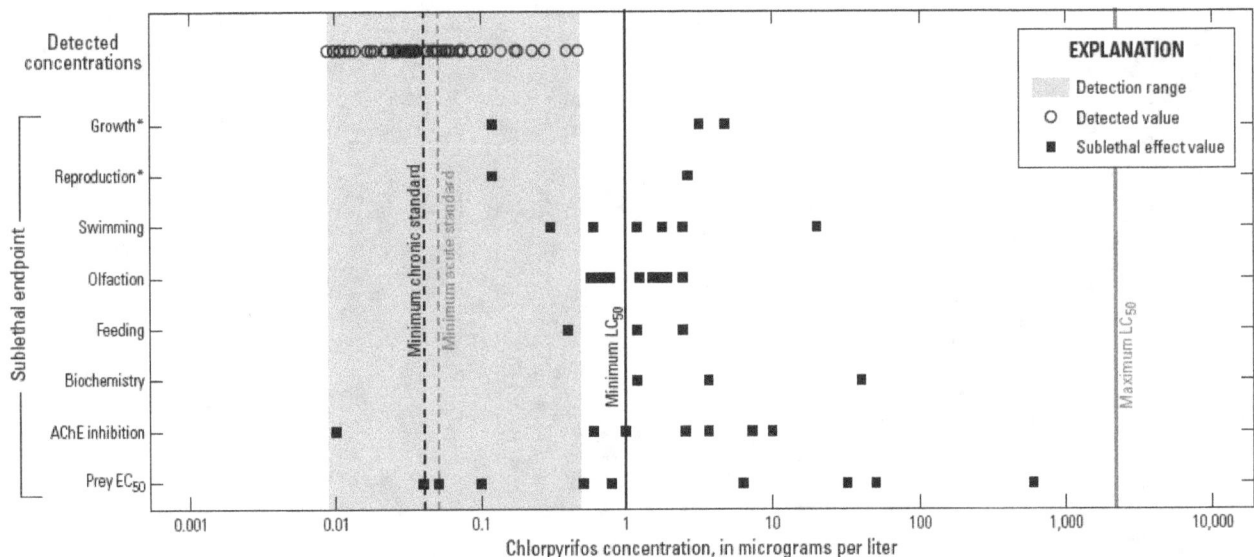

Figure 6. Detected chlorpyrifos concentrations in the Hood River basin, Oregon, compared to USEPA and Oregon water-quality standards, and toxicity and sublethal endpoints for salmonids and their prey. Acute, 24-hour exposure; chronic, 96-hour exposure; LC_{50}, 50 percent lethal concentration; EC_{50}, 50 percent effective concentration; AChE, acetylcholinesterase; *, values are for fathead minnow (*Pimephales promelas*), a less sensitive species, and are expected to be lower for salmonids.

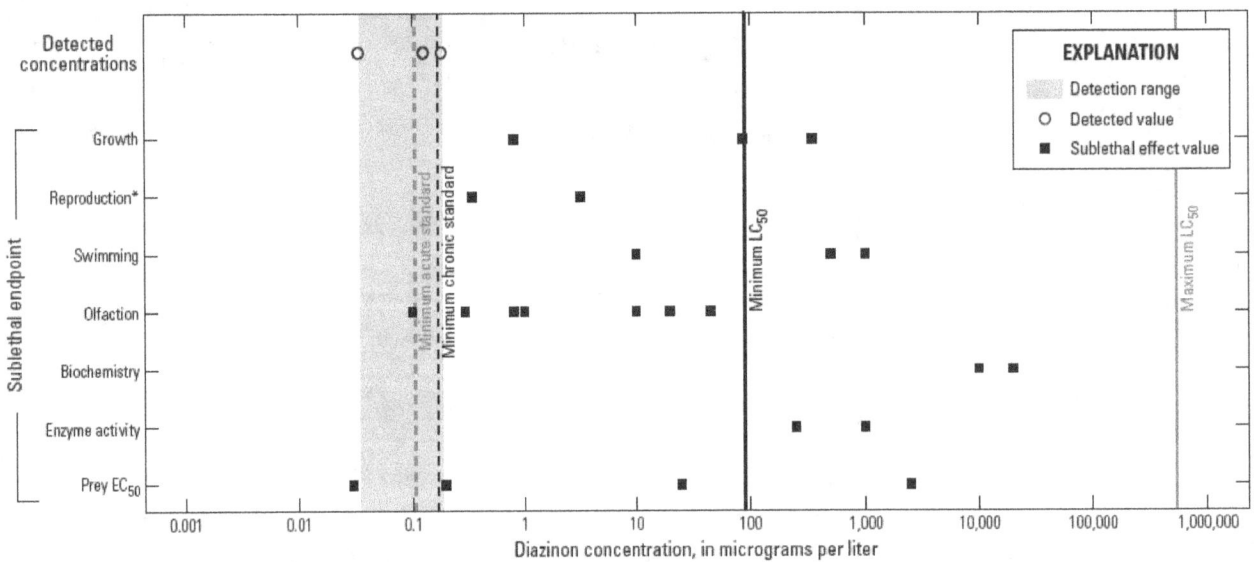

Figure 7. Detected diazinon concentrations in the Hood River basin, Oregon, compared to USEPA and Oregon water-quality standards and toxicity and sublethal endpoints for salmonids and their prey. Acute, 24-hour exposure; chronic, 96-hour exposure; LC_{50}, 50 percent lethal concentration; EC_{50}, 50 percent effective concentration; *, values are for fathead minnow (*Pimephales promelas*), a less sensitive species, and are expected to be lower for salmonids.

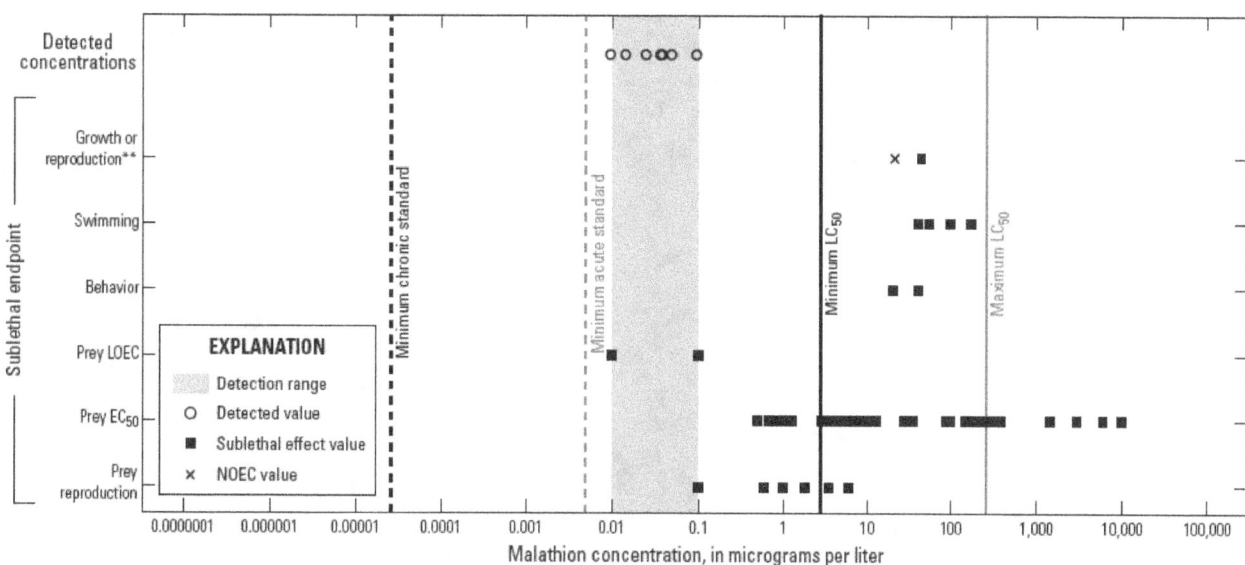

Figure 8. Detected malathion concentrations in the Hood River basin, Oregon, compared to USEPA and Oregon water-quality standards and toxicity and sublethal endpoints for salmonids and their prey. Acute, 24-hour exposure; chronic, 96-hour exposure; LC_{50}, 50 percent lethal concentration; NOEC, no observed effect concentration; EC_{50}, 50 percent effective concentration; LOEC, lowest observed effect concentration; *, values are for fathead minnow (*Pimephales promelas*), a less sensitive species, and are expected to be lower for salmonids.

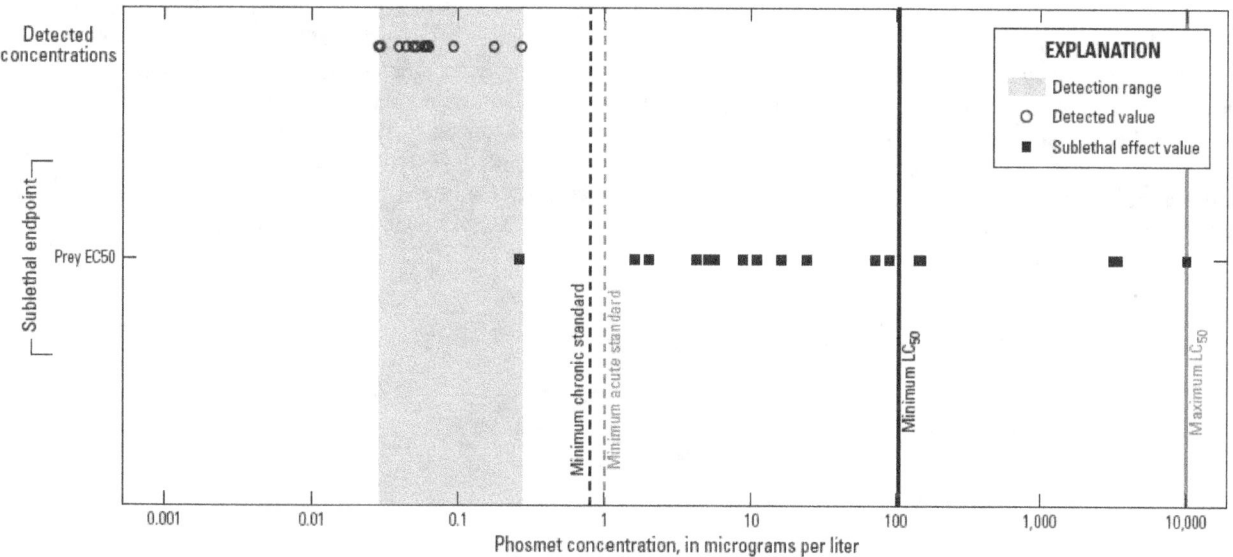

Figure 9. Detected phosmet concentrations in the Hood River basin, Oregon, compared to USEPA water-quality standards and toxicity and sublethal endpoints for salmonids and their prey. Acute, 24-hour exposure; chronic, 96-hour exposure; LC_{50}, 50 percent lethal concentration; EC_{50}, 50 percent effective concentration.

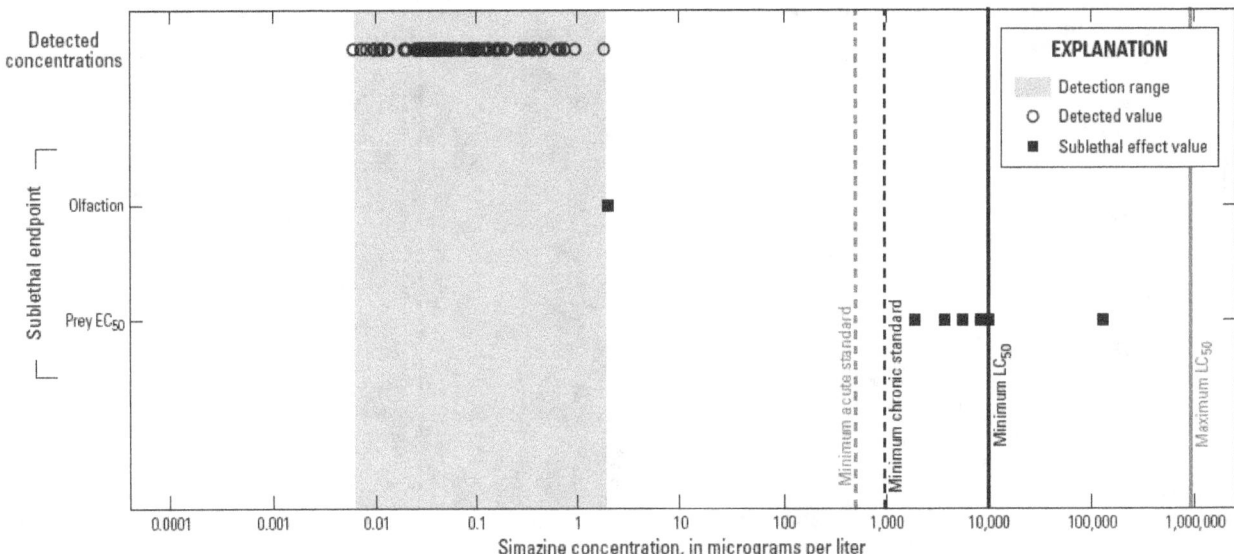

Figure 10.   Detected simazine concentrations in the Hood River basin, Oregon, compared to USEPA water-quality standards and toxicity and sublethal endpoints for salmonids and their prey. Acute, 24-hour exposure; chronic, 96-hour exposure; LC$_{50}$, 50 percent lethal concentration; EC$_{50}$, 50 percent effective concentration.

Pesticide Occurrence for Pesticides Analyzed Only in 2009

In 2009, the number of pesticides analyzed increased to 100, compared with 10 in 2008. Because of the increase in the number of analytes, 12 pesticides were detected in 2009 that had not previously been detected in the Hood River basin: 2 fungicides, 4 herbicides, and 6 insecticides. Pesticides were detected at seven of eight sites. No pesticides were detected at West Fork Hood River at Moving Falls (RM 2.5);

however, only one sample was collected at this site. All other sites were sampled 15 or 16 times in 2009. Table 7 provides sample counts, detection counts, and maximum detected concentrations for each pesticide detected in 2009 only.

Of the new pesticides detected in 2009, only one, endrin, was detected at a concentration exceeding the USEPA or Oregon water-quality standards (table 8). However, the standards are based on specific exposure durations (acute=24 hours, chronic=96 hours), so point-in-time samples are not directly comparable to the standards.

Table 7. Counts of samples and detections for pesticides in the Hood River basin, Oregon, analyzed only in 2009, sites where pesticides were detected, and concentration, date, and site of the maximum detected concentration.

[Sample size includes all (unscreened) samples. **Abbreviations:** µg/L, microgram per liter]

Pesticide type	Class	Pesticide	Number of samples	Number of detections	Reporting limit range (µg/l)	Sites where detected	Maximum concentration 2009		
							Measure-ment (µg/L)	Date	Site
Fungicide	Triazole	Propicon-azole	97	6	0.018 - 0.023	Neal, middle; Neal, mouth; Neal, upper, below diversion	0.081	10-15-09	Neal, upper, below diversion
	Strobilurin	Pyraclos-trobin	104	7	0.0038 - 0.0045	Hood, mouth; Hood, West Fork, mouth; Lenz; Neal, middle; Neal, mouth	0.071	04-21-09	Lenz
Herbicide	Substituted urea	Diuron	104	41	0.0037 - 0.0045	Hood, mouth; Hood, West Fork, mouth; Lenz; Neal, middle; Neal, mouth; Neal, upper, below diversion; Rogers	1.68	05-07-09	Lenz
		Fluometuron	104	1	0.0037 - 0.0045	Neal, upper, below diversion	0.004	04-01-09	Neal, upper, below diversion
	Triazine	Hexazinone	111	34	0.0036 - 0.0045	Lenz; Neal, middle; Neal, mouth; Neal, upper, below diversion	0.095	04-29-09	Neal, upper, below diversion
	Fluorinated pyridazinone	Norflurazon	111	1	0.018 - 0.023	Lenz	0.046	03-17-09	Lenz
Insecticide	Carbamate	Carbaryl	104	9	0.0045 - 0.0057	Lenz; Neal, middle; Neal, mouth	0.037	05-07-09	Lenz
		Methomyl	104	4	0.0018 - 0.0023	Lenz; Neal, middle; Neal, upper, below diversion; Rogers	0.005	04-29-09	Neal, upper, below diversion
		Propoxur	104	4	0.0018 - 0.0023	Lenz; Neal, middle; Neal, upper, below diversion; Rogers	0.005	04-29-09	Neal, upper, below diversion
	Chloronicotinyl	Imidacloprid	94	1	0.018 - 0.023	Lenz	0.076	10-14-09	Lenz
	N,N-dialkyl-amide	DEET	104	2	0.0045 - 0.0057	Hood, West Fork, mouth; Lenz	0.013	04-29-09	Lenz
	Organochlorine	Endrin	111	1	0.054 - 0.068	Rogers	0.080	04-29-09	Rogers

Table 8.   Percentage of samples from the Hood River basin, Oregon, exceeding U.S. Environmental Protection Agency or Oregon freshwater aquatic life standards for pesticides analyzed only in 2009.

[Sample size includes screened samples, not total samples. All concentrations reported in micrograms per liter; **Abbreviations:** USEPA, U.S. Environmental Protection Agency; CMC, criteria maximum concentration; CCC, criterion continuous concentration; %, percent; –, no water-quality standard; >, greater than]

	USEPA Office of Pesticide Programs Aquatic Life Benchmarks				USEPA Water Quality Criteria		Oregon Water Quality Criteria		Percentage exceeding lowest standard (%)
	Fish		Invertebrates						
	Acute	Chronic	Acute	Chronic	CMC (Acute)	CCC (Chronic)	Acute	Chronic	
Propiconazole	425	95	2,400	–	–	–	–	–	0
Pyraclostrobin	–	–	–	–	–	–	–	–	–
Diuron	355	26	80	160	–	–	–	–	0
Fluometuron	320	–	110	–	–	–	–	–	0
Hexazinone	137,000	17,000	75,800	20,000	–	–	–	–	0
Norflurazon	4,050	770	> 750	1,000	–	–	–	–	0
Carbaryl	110	6.8	0.85	0.5	–	–	–	–	0
DEET	–	–	–	–	–	–	–	–	–
Endrin	–	–	–	–	0.086	0.036	0.18	0.0023	0.9
Site where Rogers detected									7
Imidacloprid	> 41,500	1,200	35	1.05	–	–	–	–	0
Methomyl	160	12	2.5	0.7	–	–	–	–	0
Propoxur	1,850	–	5.5	–	–	–	–	–	0

Occurrence, Detection Frequency, and Potential Impacts on Aquatic Life for Each Detected Pesticide

Additional information on the uses and environmental fate of each pesticide is provided in appendix C and appendix E.

Carbaryl (insecticide)

Carbaryl was detected a total of nine times in 2009 in Middle Neal Creek and at the mouths of Neal Creek, Lenz Creek, and Hood River. Eight of the detections occurred in May or June; the other was in October. The highest concentration of carbaryl was more than an order of magnitude less than the lowest USEPA benchmark (0.5 µg/L) and the lowest available ecotoxicological value (1.5 µg/L, for invertebrate reproduction) (fig. 11).

DEET (insecticide)

DEET was detected twice in 2009, once each at West Fork Hood River at mouth in March and Lenz Creek at mouth in April. The detections were nearly seven orders of magnitude less than the acute toxicity values for salmonids (fig. 12). Water-quality standards do not exist for DEET.

Diuron (herbicide)

Diuron was the most frequently detected pesticide in 2009, found in 41 of 104 samples. All detected concentrations were less than the lowest aquatic life benchmarks set by the USEPA (fig. 13). The highest concentrations of diuron were found in Lenz Creek at mouth, followed by Neal Creek at mouth. Concentrations peaked in May at both sites.

Endrin (insecticide)

Endrin was detected once in Rogers Spring Creek in April 2009 at a concentration approximately 1.5 orders of magnitude higher than the ODEQ chronic water-quality criterion and within an order of magnitude less than the USEPA acute water-quality criterion and salmonid toxicity values (fig. 14). Endrin has not been registered for use since 1991, but it is very persistent in soils.

Fluometuron (herbicide)

Fluometuron was detected once in Upper Neal Creek in April 2009 at a concentration four orders of magnitude less than the lowest USEPA benchmark (fig. 15). However, the detection was unexpected given that fluometuron is registered for use only on cotton, a crop not grown in Oregon.

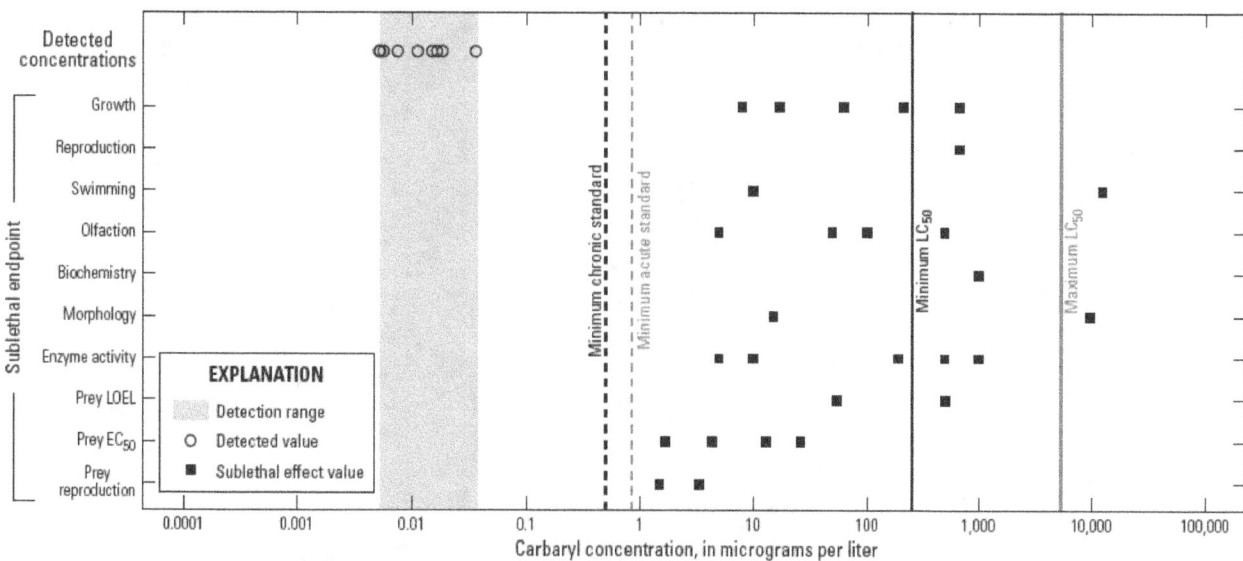

Figure 11. Detected carbaryl concentrations in the Hood River basin, Oregon, compared to USEPA and Oregon water-quality standards and toxicity and sublethal endpoints for salmonids and their prey. Acute, 24-hour exposure; chronic, 96-hour exposure; LC$_{50}$, 50 percent lethal concentration; EC$_{50}$, 50 percent effective concentration; LOEL, lowest observed effect level.

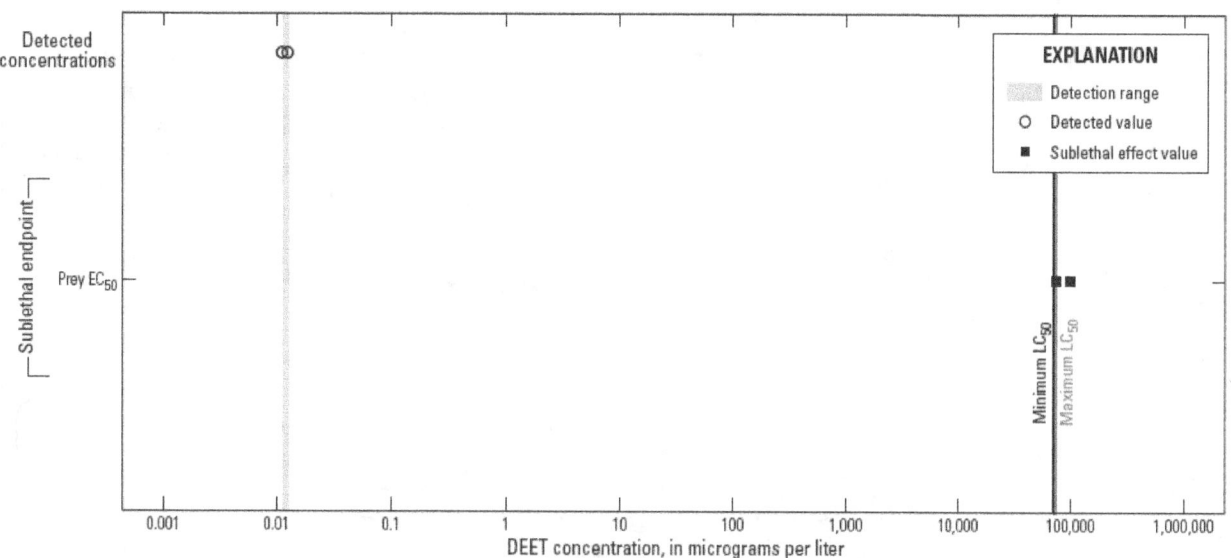

Figure 12. Detected DEET concentrations in the Hood River basin, Oregon, compared to toxicity and sublethal endpoints for salmonids and their prey. Acute, 24-hour exposure; chronic, 96-hour exposure; LC$_{50}$, 50 percent lethal concentration; EC$_{50}$, 50 percent effective concentration.

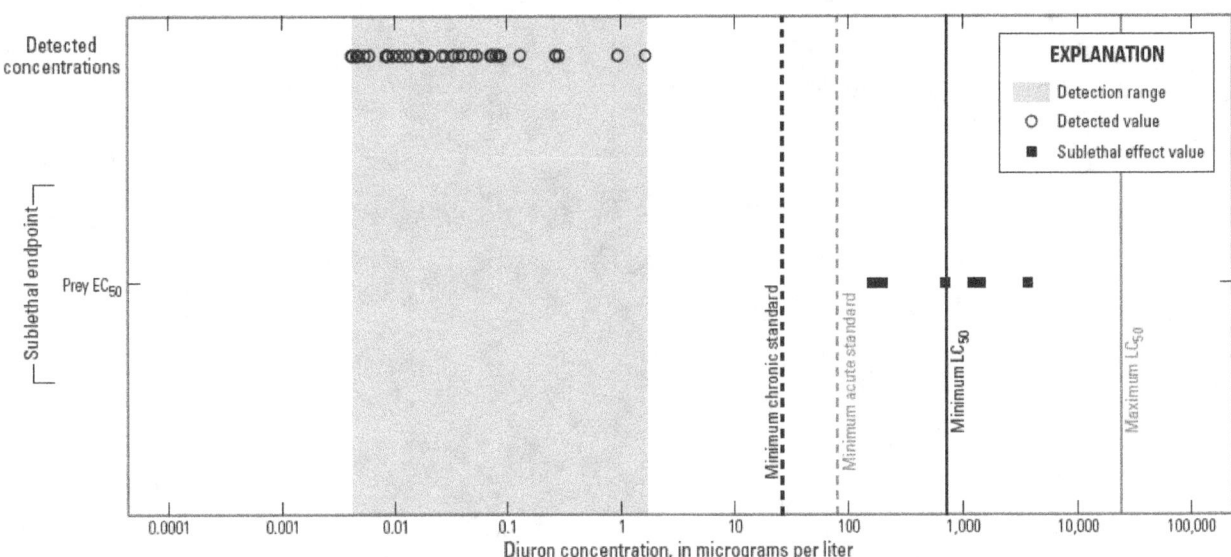

Figure 13.   Detected diuron concentrations in the Hood River basin, Oregon, compared to USEPA and Oregon water-quality standards and toxicity and sublethal endpoints for salmonids and their prey. Acute, 24-hour exposure; chronic, 96-hour exposure; LC$_{50}$, 50 percent lethal concentration; EC$_{50}$, 50 percent effective concentration.

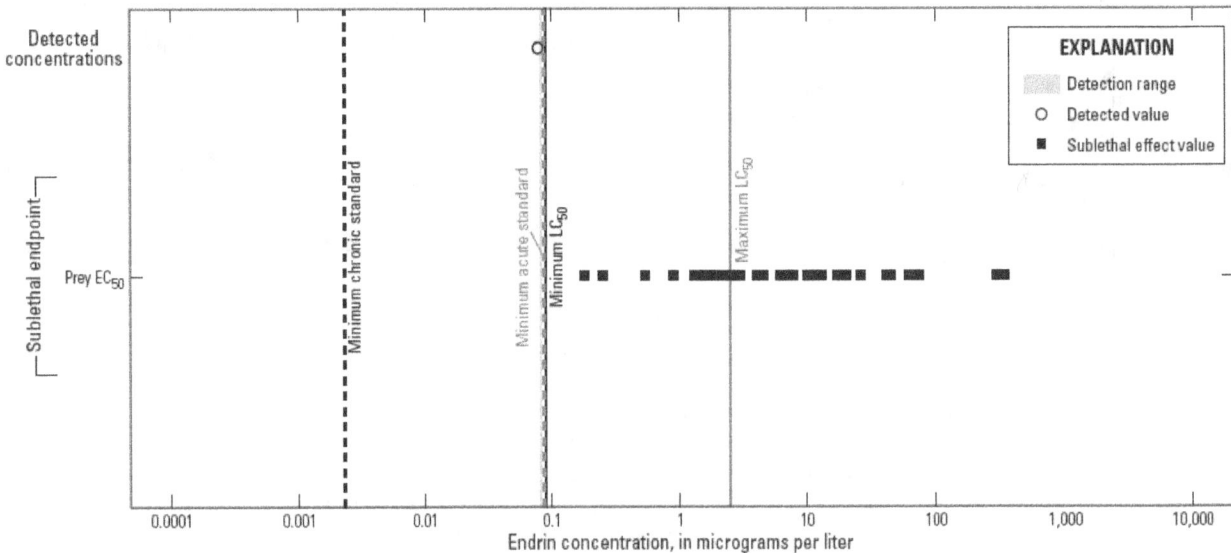

Figure 14.   Detected endrin concentrations in the Hood River basin, Oregon, compared to USEPA and Oregon water-quality standards and toxicity and sublethal endpoints for salmonids and their prey. Acute, 24-hour exposure; chronic, 96-hour exposure; LC$_{50}$, 50 percent lethal concentration; EC$_{50}$, 50 percent effective concentration.

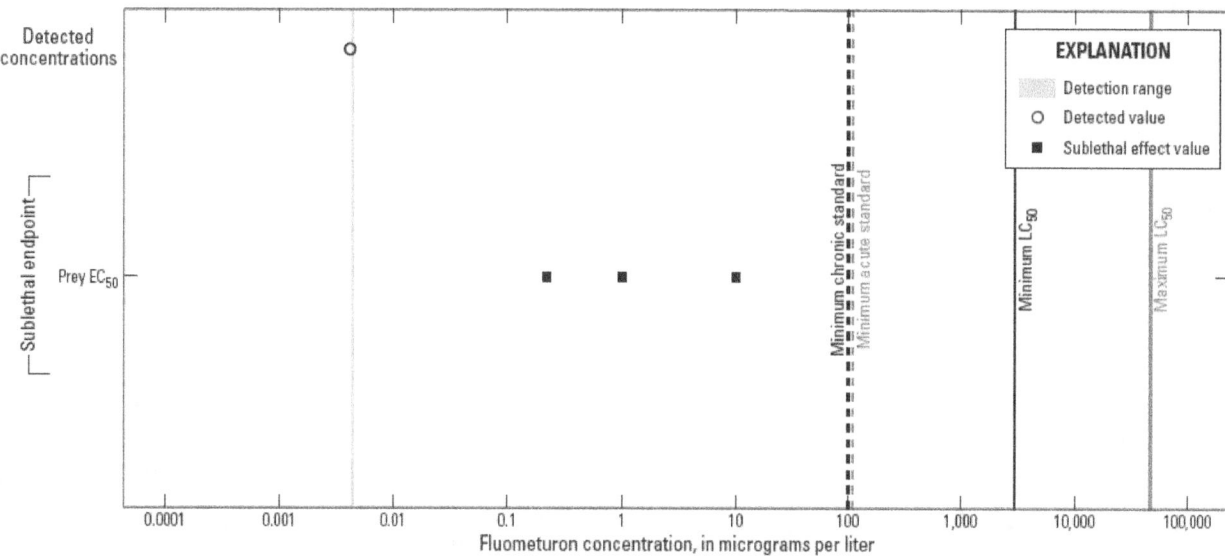

Figure 15. Detected fluometuron concentrations in the Hood River basin, Oregon, compared to USEPA water-quality standards and toxicity and sublethal endpoints for salmonids and their prey. Acute, 24-hour exposure; chronic, 96-hour exposure; LC_{50}, 50 percent lethal concentration; EC_{50}, 50 percent effective concentration.

Hexazinone (herbicide)

Hexazinone was detected in 34 of 111 samples in March through June 2009, with the highest concentrations occurring in April and May. All detections except one were from the three Neal Creek sites (Upper below agricultural diversion, Middle, and at mouth). The other detection was in Lenz Creek at mouth. The highest concentration of hexazinone found was more than five orders of magnitude lower than the lowest USEPA benchmark (fig. 16). Hexazinone is a broad-range triazine herbicide, predominantly used in forestry, the dominant upstream land use in the Neal Creek basin. While it is not highly toxic to salmonids, it can harm salmonid habitat by affecting vegetation as far as 100 meters from the application site because of its persistence and mobility in soil and surface waters (Wan and others, 1988).

Imidacloprid (insecticide)

Imidacloprid was detected once at Lenz Creek at mouth in October 2009. Its concentration was more than an order of magnitude lower than the lowest USEPA benchmark (fig. 17).

Methomyl (insecticide)

Methomyl was detected at four sites on April 29, 2009: at Upper Neal and Rogers Spring Creeks, and at more dilute concentrations at Middle Neal Creek and at Lenz Creek at mouth. All detections were several orders of magnitude lower than USEPA benchmarks or concentrations known to induce sublethal responses in salmonids or their prey (fig. 18).

Norflurazon (herbicide)

Norflurazon was detected once in March 2009 in Lenz Creek at mouth at a concentration several orders of magnitude less than salmonid toxicity endpoints and USEPA benchmarks (fig. 19).

Propiconazole (fungicide)

Propiconazole was detected six times in 2009; all concentrations were much less than the lowest water-quality benchmarks (95 and 425 µg/L) (fig. 20). It was detected at the three Neal Creek sites (Upper below agricultural diversion, Middle, and at mouth) in March, May, and October, with generally higher concentrations detected at the upstream sites.

Propoxur (insecticide)

Propoxur was detected in Middle Neal, Upper Neal, Lenz, and Rogers Spring Creeks on April 29, 2009. Detected concentrations were several orders of magnitude less than those known to harm salmonids or their prey (fig. 21).

Pyraclostrobin (fungicide)

Pyraclostrobin was detected a total of seven times in 2009: at the mouths of Neal Creek, Lenz Creek, Hood River, and West Fork Hood River and in Middle Neal Creek. Detections occurred in April, except for one in May at Lenz Creek at mouth. All detected concentrations were less than acute toxicity values for salmonids and invertebrates (fig. 22). No State or USEPA benchmarks exist for pyraclostrobin.

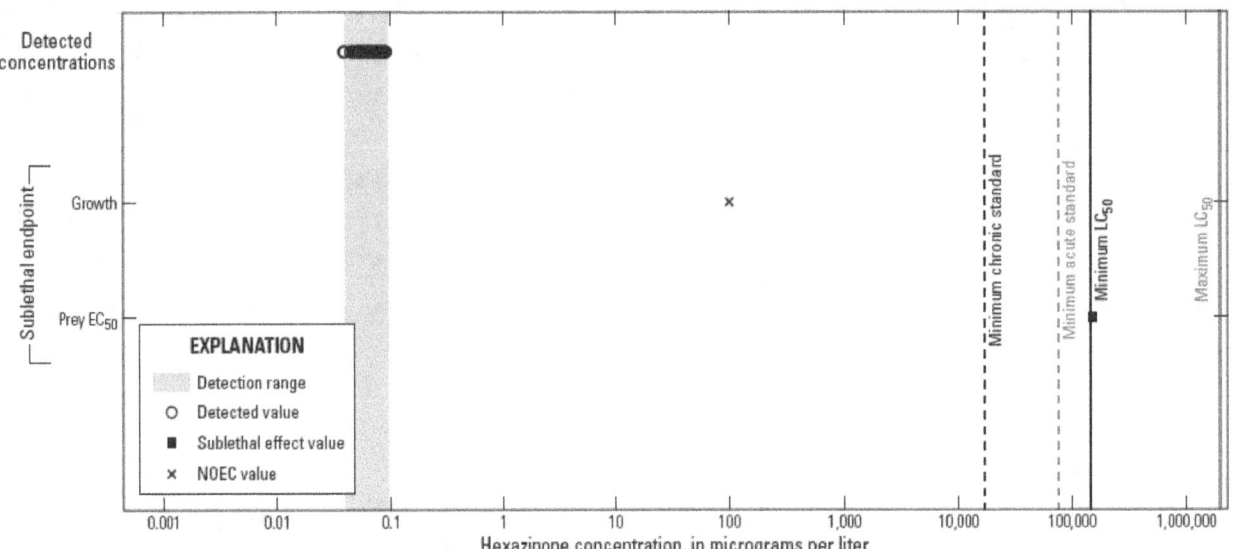

Figure 16.   Detected hexazinone concentrations in the Hood River basin, Oregon, compared to USEPA water-quality standards and toxicity and sublethal endpoints for salmonids and their prey. Acute, 24-hour exposure; chronic, 96-hour exposure; LC$_{50}$, 50 percent lethal concentration; EC$_{50}$, 50 percent effective concentration; NOEC, no observed effect concentration.

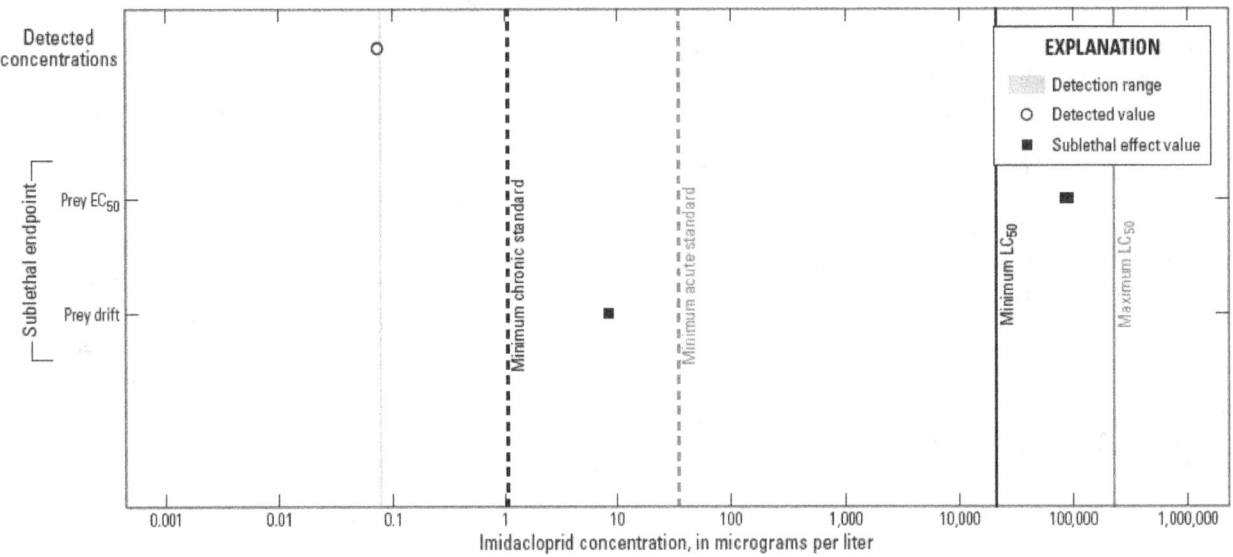

Figure 17.   Detected imidacloprid concentrations in the Hood River basin, Oregon, compared to USEPA water-quality standards and toxicity and sublethal endpoints for salmonids and their prey. Acute, 24-hour exposure; chronic, 96-hour exposure; LC$_{50}$, 50 percent lethal concentration; EC$_{50}$, 50 percent effective concentration.

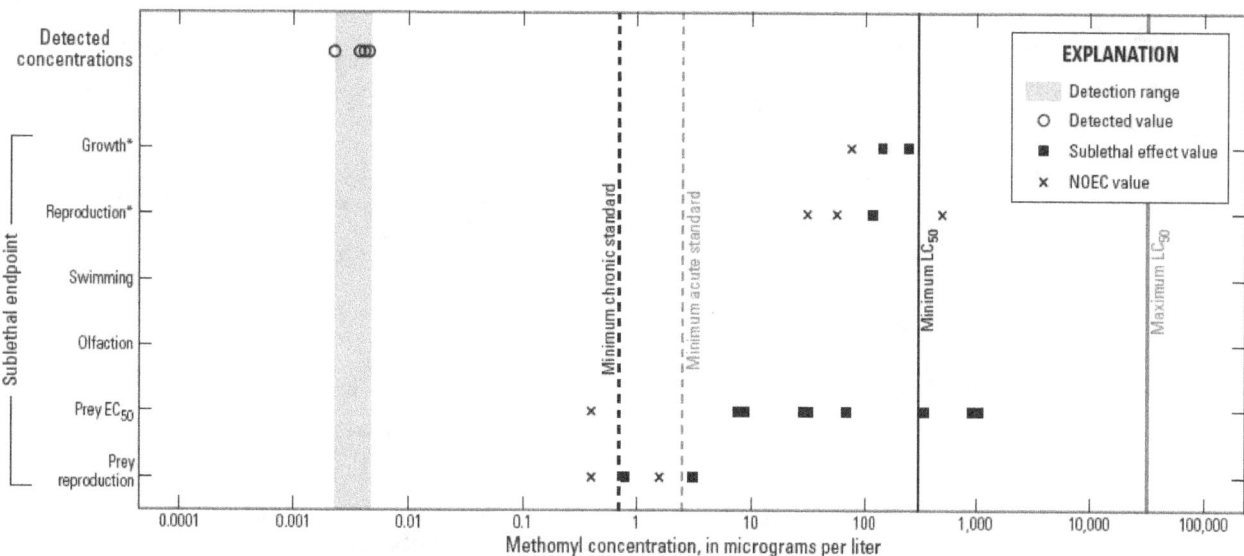

Figure 18. Detected methomyl concentrations in the Hood River basin, Oregon, compared to USEPA water-quality standards and toxicity and sublethal endpoints for salmonids and their prey. Acute, 24-hour exposure; chronic, 96-hour exposure; LC$_{50}$, 50 percent lethal concentration; NOEC, no observed effect concentration; EC$_{50}$, 50 percent effective concentration; *, values are for fathead minnow (*Pimephales promelas*), a less sensitive species, and are expected to be lower for salmonids.

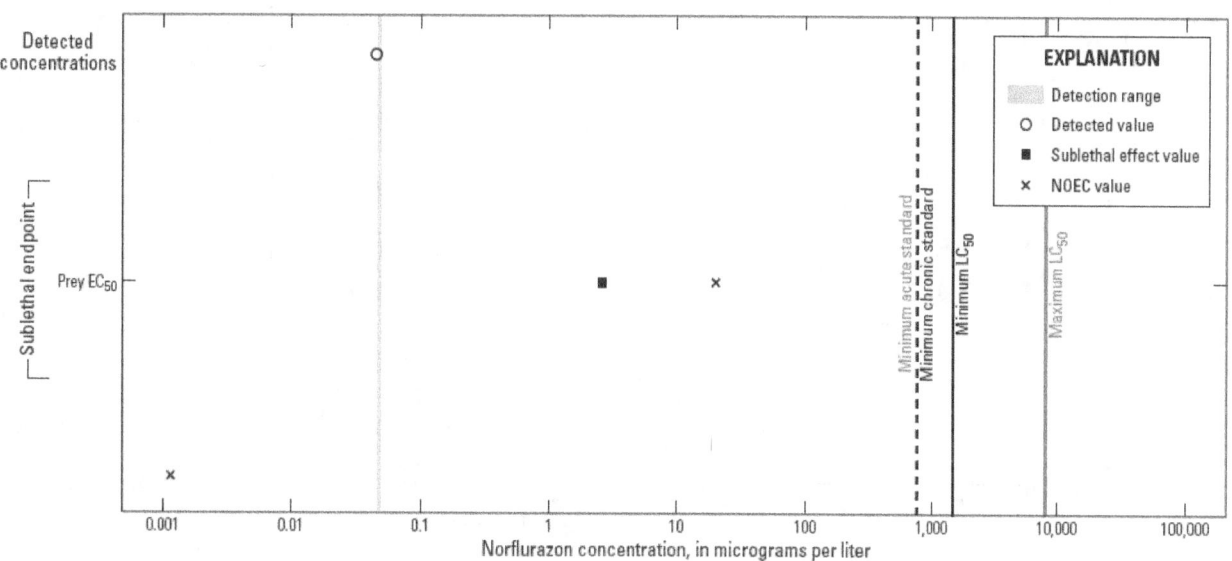

Figure 19. Detected norflurazon concentrations in the Hood River basin, Oregon, compared to USEPA water-quality standards and toxicity and sublethal endpoints for salmonids and their prey. Acute, 24-hour exposure; chronic, 96-hour exposure; LC$_{50}$, 50 percent lethal concentration; NOEC, no observed effect concentration; EC$_{50}$, 50 percent effective concentration.

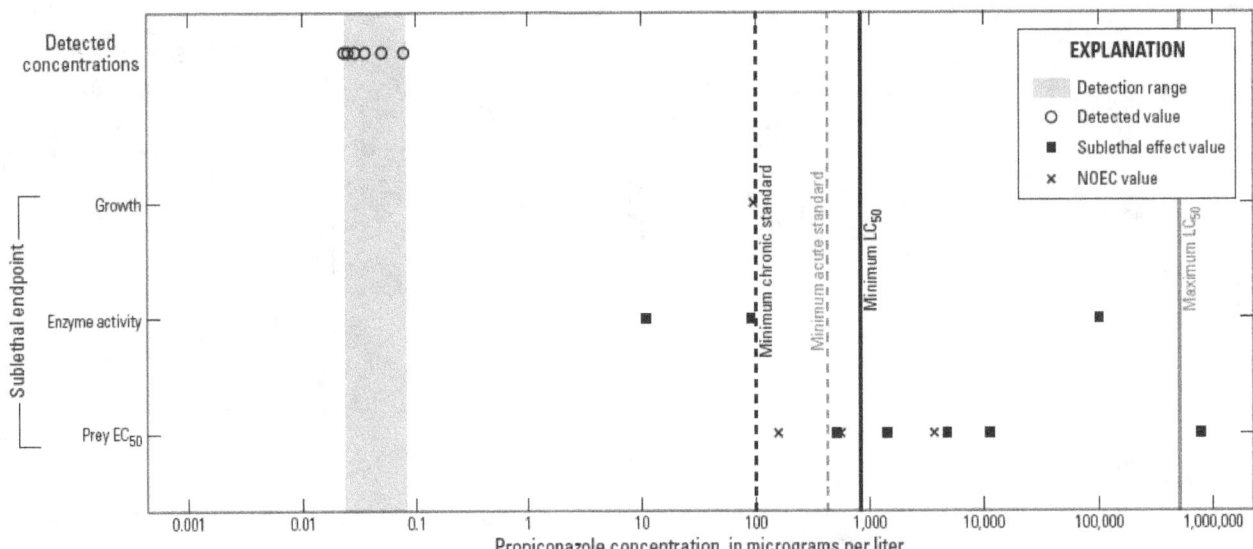

Figure 20.   Detected propiconazole concentrations in the Hood River basin, Oregon, compared to USEPA water-quality standards and toxicity and sublethal endpoints for salmonids and their prey. Acute, 24-hour exposure; chronic, 96-hour exposure; LC$_{50}$, 50 percent lethal concentration; NOEC, no observed effect concentration; EC$_{50}$, 50 percent effective concentration.

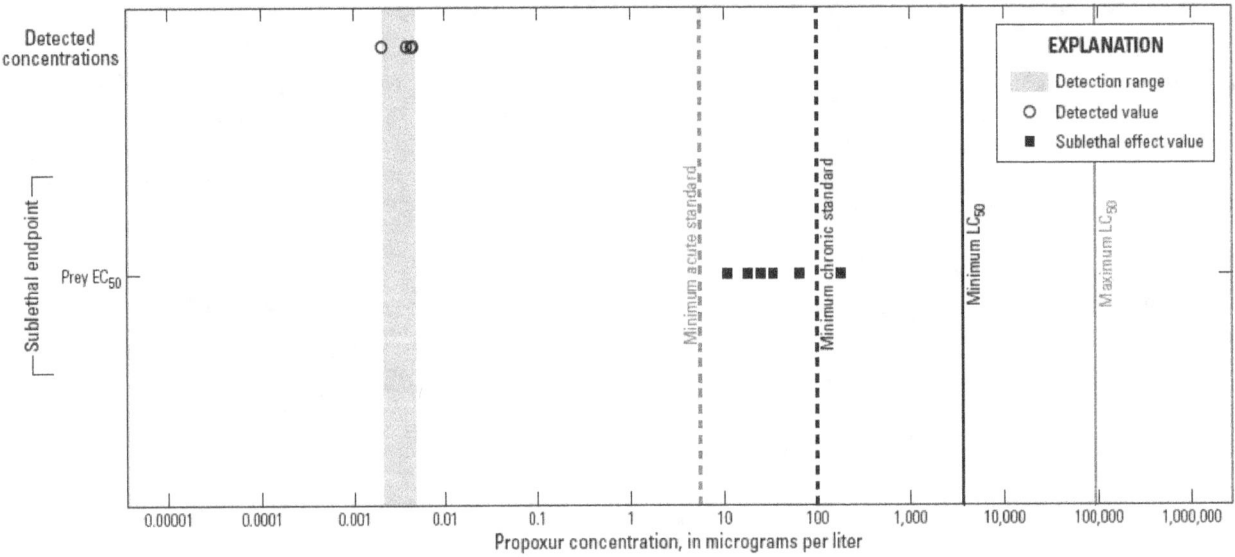

Figure 21.   Detected propoxur concentrations in the Hood River basin, Oregon, compared to USEPA benchmarks and toxicity and sublethal endpoints for salmonids and their prey. Acute, 24-hour exposure; chronic, 96-hour exposure; LC$_{50}$, 50 percent lethal concentration; EC$_{50}$, 50 percent effective concentration.

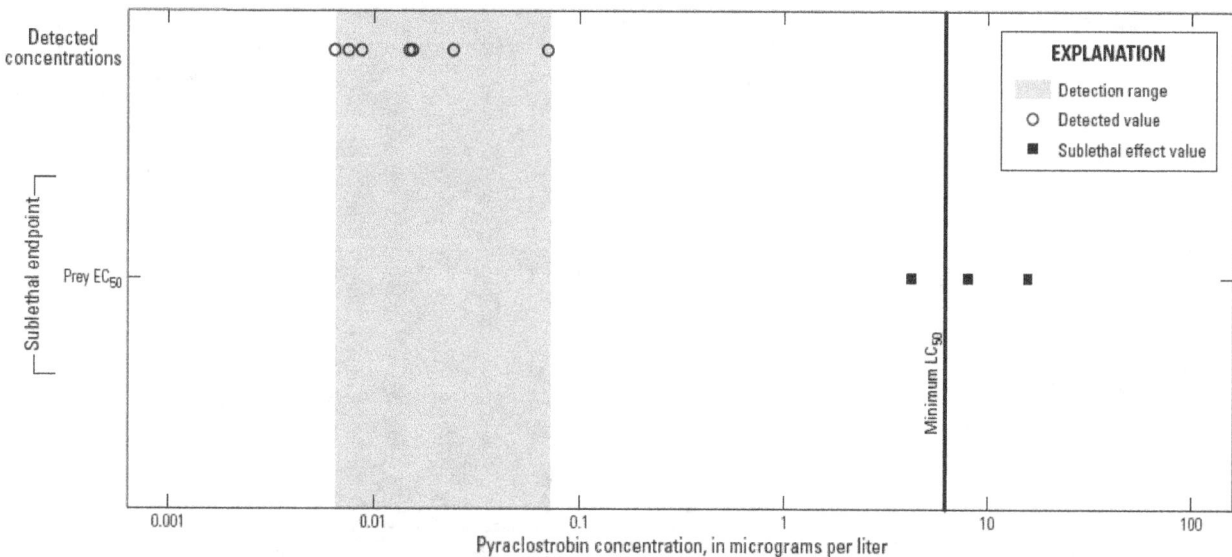

Figure 22. Detected pyraclostrobin concentrations in the Hood River basin, Oregon, compared to toxicity and sublethal endpoints for salmonids and their prey. LC$_{50}$, 50 percent lethal concentration; EC$_{50}$, 50 percent effective concentration.

Trends in Pesticide Detections and Concentrations

Subsets of the 1999–2009 dataset from the mouths of Neal Creek, Lenz Creek, and Hood River were used to examine long-term trends throughout the basin. Only one site, Neal Creek at mouth, was monitored every year from 1999 through 2009 (162 samples). Lenz Creek at mouth was monitored from 2001 through 2006 and 2008 through 2009 (114 samples). The mouth of Hood River was monitored during 1999–2002 and 2005–2009 (90 samples). Data were screened in order to draw comparisons across years with different reporting limits (refer to the Methods section for more information on data screening). Appendix G contains screened sample counts and detections by site for all pesticides.

Neal Creek at Mouth

Azinphos-methyl, chlorpyrifos, and simazine were the most frequently detected pesticides at Neal Creek at mouth. Azinphos-methyl was detected mostly in the summer and fall and has not been detected since 2007 (fig. 23). About half of azinphos-methyl samples from this site in 1999, 2008, and 2009 were removed during the data screening process due to high reporting limits. Chlorpyrifos was detected during March, except one detection in April, with the last detections in 2005 (fig. 24). However, fewer samples collected in March and April were analyzed for chlorpyrifos after 2005 compared to previous years. All chlorpyrifos data at this site for 2009 were removed from the screened dataset due to the high reporting limit. However, chlorpyrifos was not detected in 14 samples in 2009 at or exceeding the lowest water-quality criterion (0.041 µg/L). The highest detection counts and concentrations of simazine occurred during the summer months, but simazine was also detected in the spring. It was detected every year except 2008, with general downward trends in the percentage of samples with detections and in annual maximum concentration (fig. 25). From 1999 through 2006, simazine was detected in 43 percent of samples at Neal Creek at mouth. From 2007 through 2009, it was detected in 14 percent of samples.

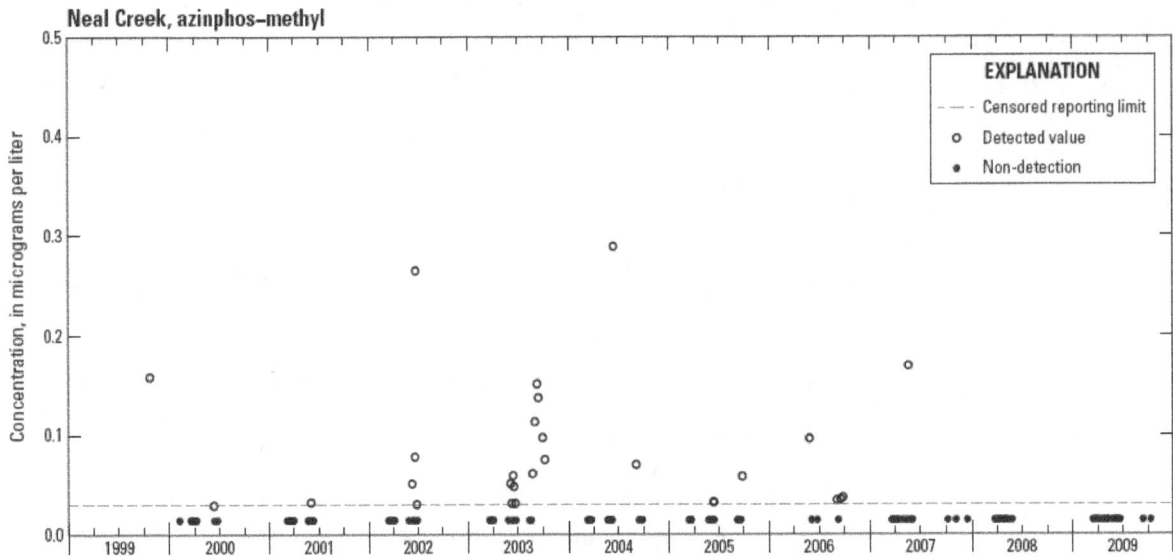

Figure 23.   Azinphos-methyl concentrations in Neal Creek at mouth, 1999–2009. Nondetections are shown at one-half the censored reporting limit.

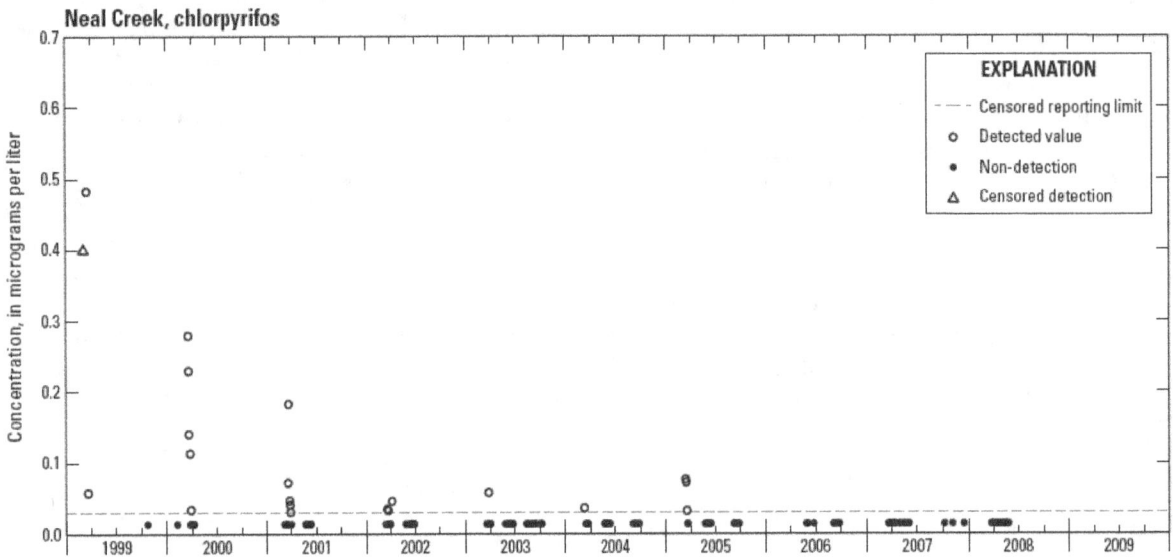

Figure 24.   Chlorpyrifos concentrations in Neal Creek at mouth, 1999–2009. Nondetections are shown at one-half the censored reporting limit. A censored detection is a detected concentration that was screened out of the dataset due to its high reporting limit.

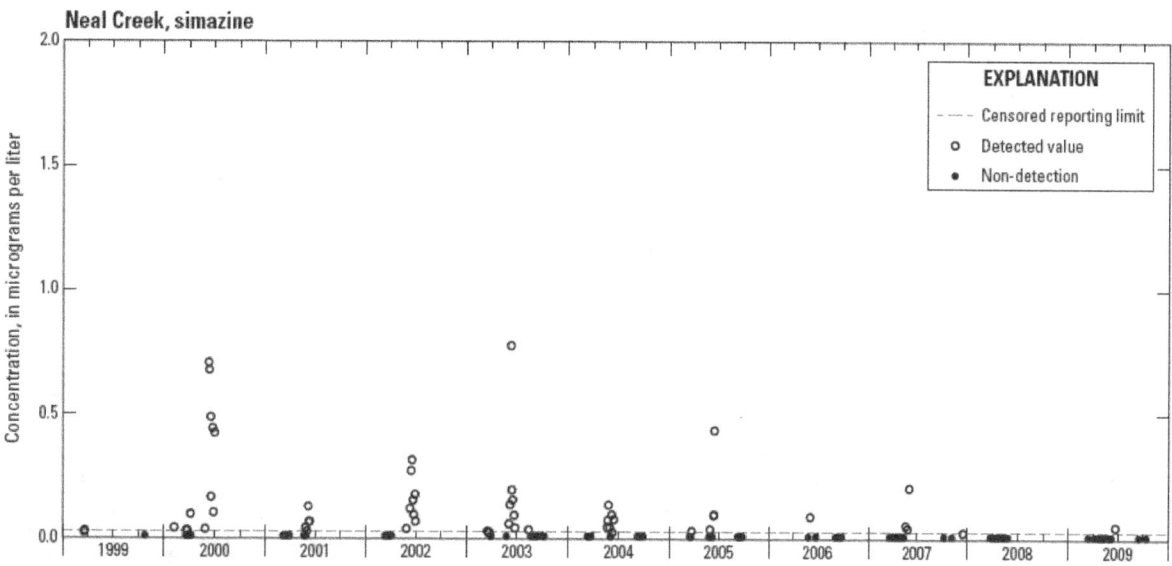

Figure 25.   Simazine concentrations in Neal Creek at mouth, 1999–2009. Nondetections are shown at one-half the censored reporting limit.

Lenz Creek at Mouth

Azinphos-methyl, chlorpyrifos, phosmet, and simazine were the most frequently detected pesticides at Lenz Creek at mouth since 2001. Azinphos-methyl detections were nearly evenly split between summer and fall. Azinphos-methyl was found in 0 to 70 percent of samples from 2001 through 2005 (fig. 26), when the annual sample counts ranged from 16 to 20. Only three samples were collected in 2006. On average, it was present in 32 percent of samples from 2001 through 2006 and 0 percent of samples since 2008 (there are no samples from 2007). However, many azinphos-methyl samples from 2008–09 were screened out.

Chlorpyrifos was detected one to four times in March during each year from 2001 through 2005 and has not been detected since (fig. 27). However, no chlorpyrifos samples were analyzed in March in 2006–2008. Moreover, there were 87 samples from 2001 to 2005 and only 10 since 2006 due to a decrease in total sample counts in 2006 and 2008, the complete absence of samples in 2007, and the censoring of 2009 data due to the high reporting limit.

Phosmet was detected one to three times per year from 2002 through 2005 (fig. 28), representing 17 percent or less of yearly samples. Overall, it was detected in 7.7 percent of samples through 2006 and in 4.3 percent of samples since 2007. It was detected mostly during the fall. The absence of samples collected during fall after 2005 may have caused the nearly complete absence of detections since then.

Simazine was most frequently detected and was found at the highest concentrations in Lenz Creek during the summer. It was detected in at least 28 percent of screened samples each year in 2001–06 and 2008–09 (fig. 29). Three to 20 samples were collected per year. Prior to 2007, simazine was detected in 61 percent of screened samples from Lenz Creek. Since 2008, it was detected in 53 percent of samples. Many simazine samples from 2008 to 2009 were removed because they had a higher reporting limit than the censoring level (0.027 µg/L).

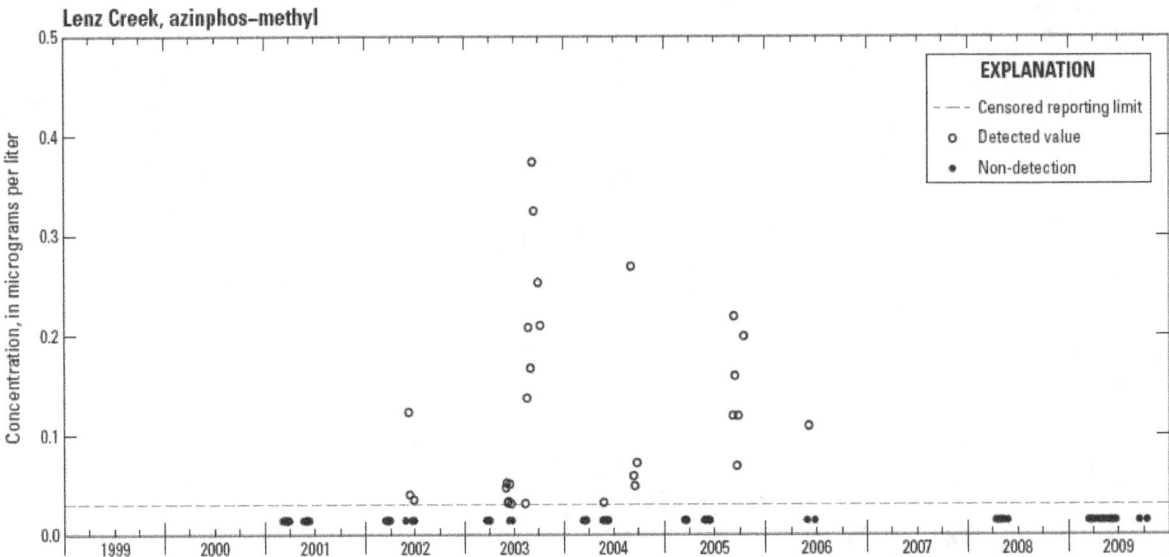

Figure 26. Azinphos-methyl concentrations in Lenz Creek at mouth, 2001–2009. Nondetections are shown at one-half the censored reporting limit.

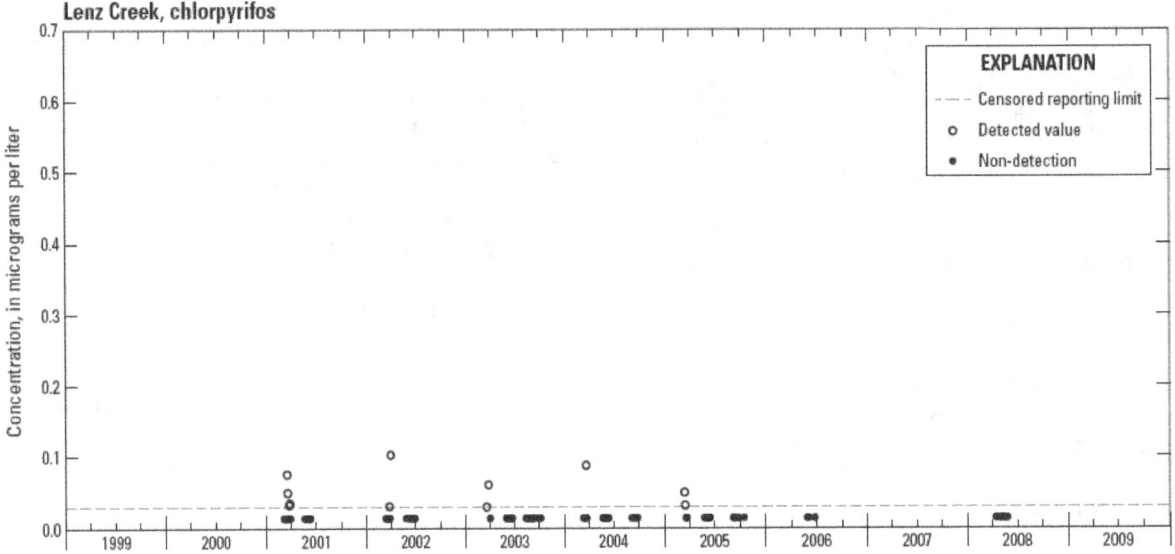

Figure 27. Chlorpyrifos concentrations in Lenz Creek at mouth, 2001–2009. Nondetections are shown at one-half the censored reporting limit.

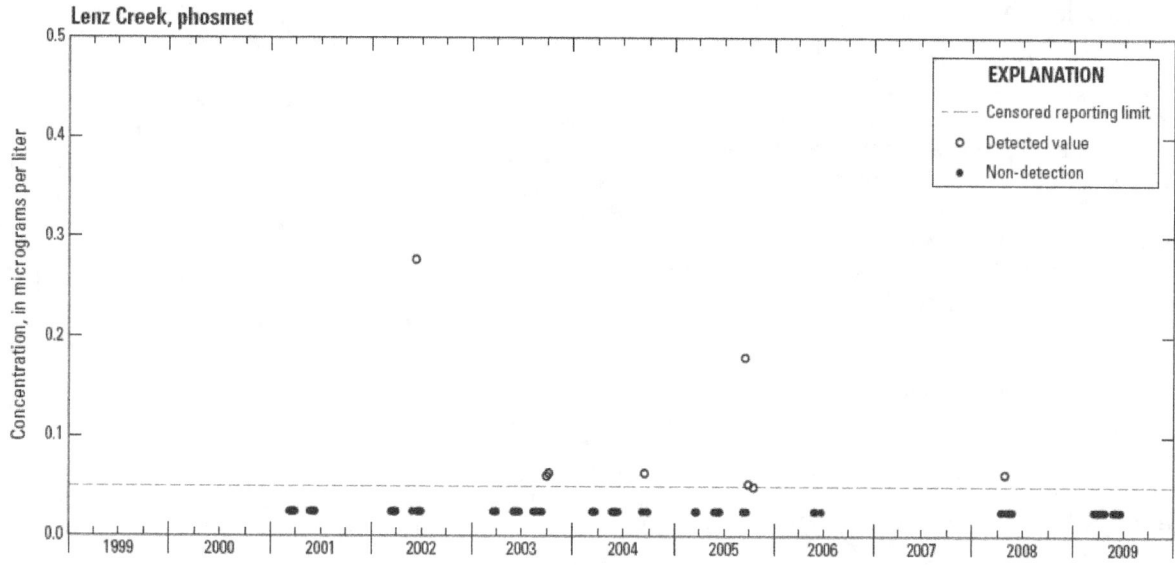

Figure 28. Phosmet concentrations in Lenz Creek at mouth, 2001–2009. Nondetections are shown at one-half the censored reporting limit.

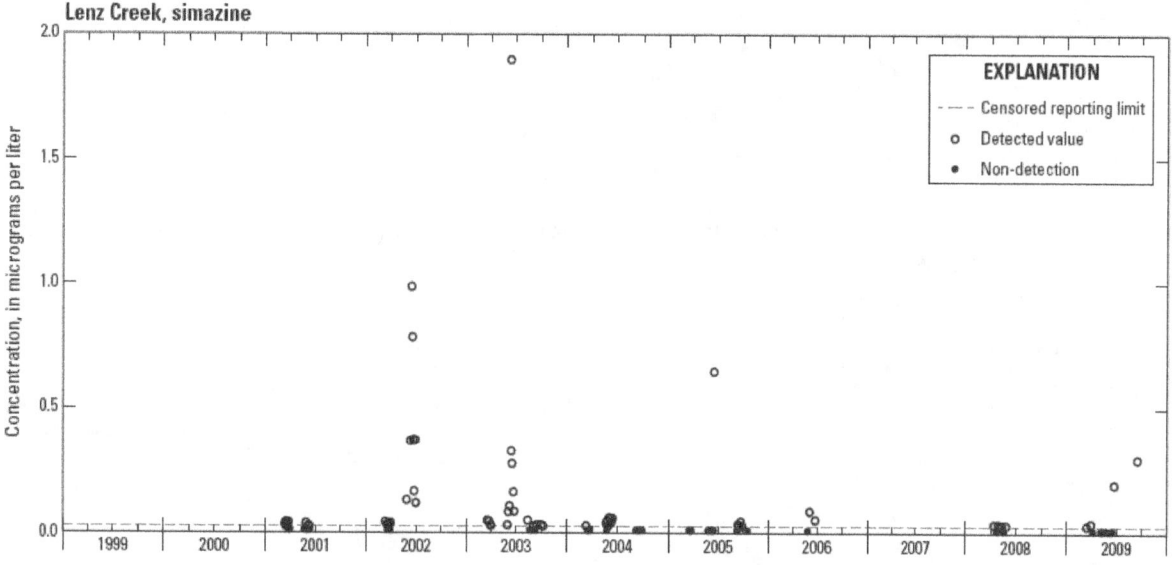

Figure 29. Simazine concentrations in Lenz Creek at mouth, 2001–2009. Nondetections are shown at one-half the censored reporting limit.

Hood River at Mouth

The mouth of Hood River was monitored from 1999 through 2002 and 2005 through 2009. Samples were collected from two sites at the mouth of Hood River: Hood River downstream of Ppl Powerdale Powerhouse (1999-2001) and Hood River at footbridge downstream of I-84 (2002 and 2005–2009). Pesticides were detected in fewer samples at the mouth of Hood River than at Lenz Creek at mouth or Neal Creek at mouth. Chlorpyrifos was detected once each in 1999 and 2001 (fig. 30). Approximately twice as many

samples were analyzed for chlorpyrifos each year after 2001 compared to before 2001, so the absence of detections was not caused by the number of samples collected. It was found in 4 percent of 45 samples through 2006 and 0 percent of 23 samples in 2007–08. Chlorpyrifos data from 2009 were screened out due to the higher reporting limit. Simazine was detected in 2 of 4 samples in 2000, 2 of 15 samples in 2005, and 2 of 9 samples in 2009 (fig. 31). Four of the six samples with detections were collected in May or June. Simazine was found in 4 percent of 47 samples through 2006 and 6 percent of 32 samples since 2007.

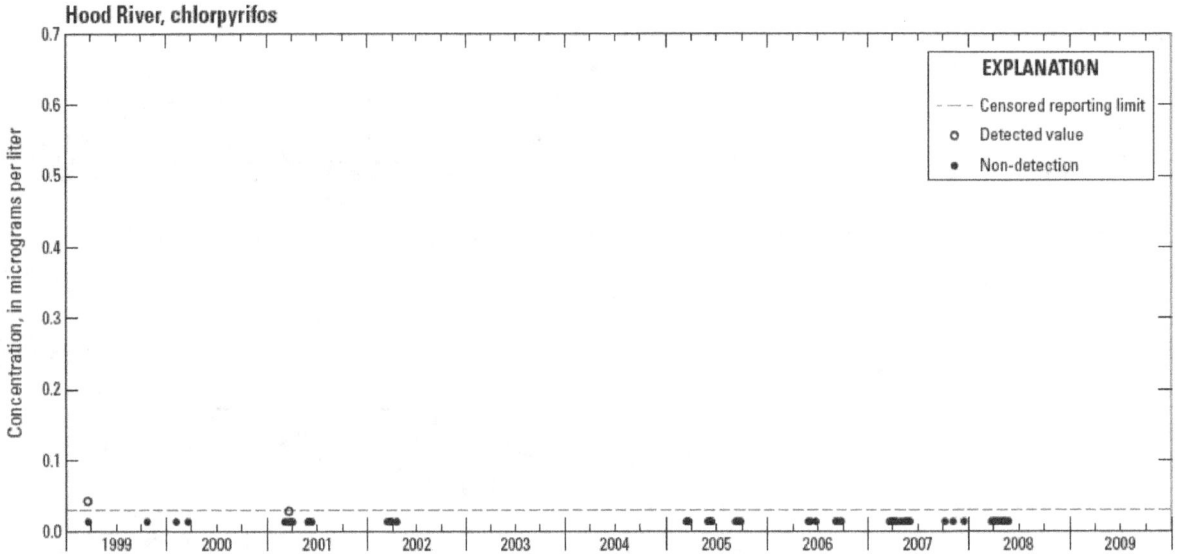

Figure 30.   Chlorpyrifos concentrations in the mouth of Hood River, 1999–2009. Nondetections are shown at one-half the censored reporting limit.

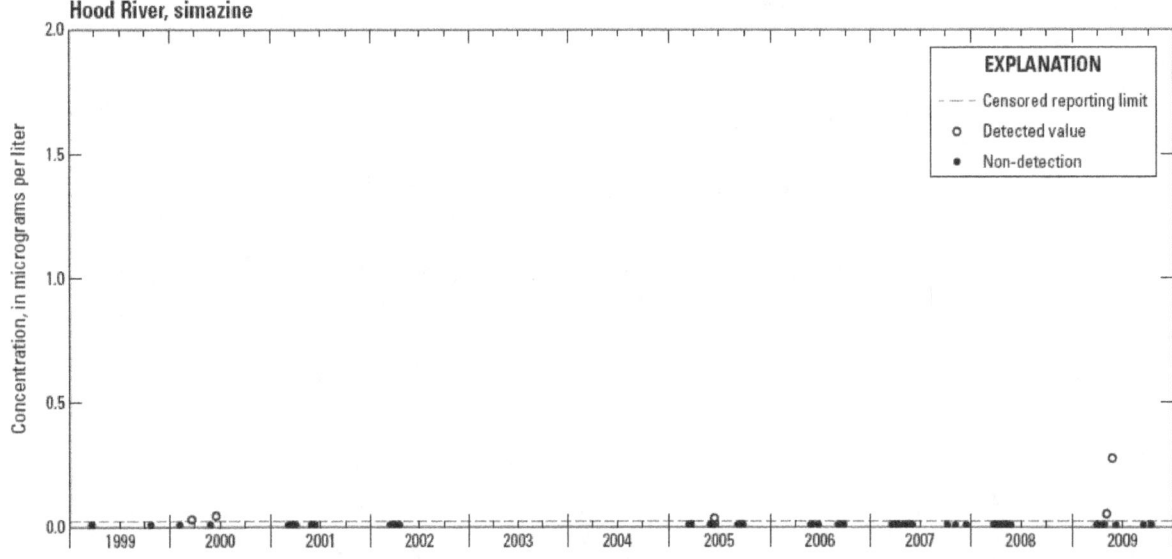

Figure 31.   Simazine concentrations in the mouth of Hood River, 1999–2009. Nondetections are shown at one-half the censored reporting limit.

Pesticide Mixtures

Of the 953 pesticide unscreened samples from 1999 through 2009, 116 (12 percent) had multiple pesticides detected in the same sample (table 9). Simazine was the pesticide most commonly detected with other pesticides in the same sample (79 percent of mixture samples). Azinphos-methyl was found in 43 percent of mixture samples, followed by diuron (31 percent), chlorpyrifos (20 percent), and hexazinone (13 percent). The most commonly detected pesticide pair was azinphos-methyl and simazine (n=47). Most pesticide mixture samples were collected from Neal Creek at mouth and Lenz Creek at mouth (45 and 38 percent, respectively). Most mixtures were detected in 2009, when the list of pesticides analyzed was much larger than in previous years. The second-highest number of pesticide mixtures was detected in 2003, the year with the highest sample count from Lenz Creek at mouth and Neal Creek at mouth. The highest number of mixture samples was collected in June (n=36). October had the highest proportion of samples collected with mixtures present, possibly driven by the low number of samples collected in October.

Pesticide Concentration Data from the Special Study on Effluent from Fruit Packers, 2004–2005

Four of 10 pesticides were detected from the 2004–05 special study on effluent from fruit packers. The number of detections and minimum, median, and maximum detected concentrations of the four detected pesticides at each site are shown in table 10. Simazine was the only pesticide that was detected at concentrations less than the lowest water-quality standard (table 11). Azinphos-methyl was the most commonly detected pesticide in these samples and it occurred in the highest concentrations, with more than half of samples exceeding the most stringent USEPA or ODEQ standard. Its degradation product, azinphos-methyl oxon, was detected in 21 of 50 samples at concentrations of 0.05–10 µg/L, with higher concentrations in fruit packing effluent and surface waters downstream of effluent discharge points. Malathion was detected once in Lenz Creek in September 2005 at a concentration exceeding the lowest acute and chronic USEPA standards, which are set to protect invertebrates (table 11). Two samples exceeded the minimum water-quality criterion for phosmet. Malathion- and phosmet-oxons were not detected.

Trace Element Concentration Data, 1999–2009

A summary of available data for priority trace elements is shown in table 12, along with USEPA and Oregon water-quality standards and values from published studies on salmonid exposure to trace elements. Table 13 shows the same information for nonpriority trace elements.

Based on total recoverable trace element concentration data from 2000 through 2002 at the nine most frequently sampled sites (n=13–39 per site), aluminum, copper, zinc, and nickel were the most commonly detected priority trace elements. The percentage of samples with those trace elements detected was highest at the mouths of Lenz Creek, Neal Creek, and Hood River and at Evans Creek at bridge. Middle Fork Hood River, West Fork Hood River, and Dog Creek had the lowest incidence of priority trace element detections from 2000 through 2002. Aluminum was detected in 100 percent of samples at all frequently sampled sites except Middle Fork Hood River and West Fork Hood River. Total recoverable copper and nickel were frequently detected at Evans Creek at bridge and the mouths of Hood River, Lenz Creek, and Neal Creek. Copper was also commonly detected at East Fork Hood River, and nickel was frequently detected at Upper Neal Creek. Zinc was detected with the same spatial pattern as nickel, but generally with lower detection frequencies. The sources of these elements in the environment could not be assessed from the available data.

Table 9. Occurrence of multiple pesticides in a single sample from Hood River basin, Oregon, 1999–2009.

[Numbers in parentheses (n) indicate the number of samples in which that combination was detected. Where not listed, n=1. Sample size includes all (unscreened) samples. **Abbreviations:** AT, atrazine; AZ, azinphos-methyl; C, chlorpyrifos; CB, carbaryl; DR, diuron; EN, endrin; FL, fluometuron; HX, hexazinone; IM, imidacloprid; MA, malathion; MM, methomyl; N, norflurazon; PH, phosmet; PX, propoxur; PZ, propiconazole; PY, pyraclostrobin; S, simazine; – no mixtures detected]

Site	Year	Mar	Apr	May	Jun	Jul	Aug	Sep	Oct	Annual Total
Evans	2003	C,S	–	–	–	–	–	–	–	1
Hood, mouth	2000	C,S	–	–	–	–	–	–	–	1
	2009	–	–	CB,S DR,S	–	–	–	–	–	2
Hood, West Fork, mouth	2009	–	–	DR,S	–	–	–	–	–	1
Indian	1999	C,S (n=2)	–	–	–	–	–	–	–	2
Lenz	2001	C,S (n=2)	–	–	–	–	–	–	–	2
	2002	C,S	C,S	AT,S	AZ,PH,S AZ,S	AZ,S	–	–	–	6
	2003	C,S AT,C,S	–	–	AZ,S (n=7)	–	AZ,S (n=2)	AZ,S (n=2)	AZ,PH, S (n=2)	15
	2004	AZ,S	–	AZ,S	AZ,MA,S	–	–	AZ,PH	–	4
	2005	–	–	–	–	–	–	AZ,PH,S (n=2) AZ,S (n=2)	AZ,PH	5
	2006	–	–	–	AZ,S	–	–	–	–	1
	2008	–	PH,S	–	–	–	–	–	–	1
	2009	DR,N,S	DR,PY DR,S (n=2) PX,DR,MM,PY	CB,DR CB,DR,HX,PY	DR,S	–	–	AZ,DR,S	CB,DR,IM	10
Neal, middle	2005	–	–	–	–	–	–	AZ,PH AZ,PH,S	–	2
	2007	–	–	PH,S	–	–	–	–	–	1
	2009	–	HX,PY PX,HX,MM	DR,HX DR,HX,PZ,S	DR,HX	–	–	–	DR,PZ	6
Neal, mouth	1999	C,S, (n=2)	–	–	–	–	–	–	AZ,S	3
	2000	C,S, (n=5)	C,S (n=4)	CB,DR,S	AZ,CB,DR,MA,S AZ,CB,DR,S AZ,DR,MA,S CB,DZ,DR,S DR,S (n=2)	CB,DR,MA,S	–	–	–	17
	2001	–	–	–	AZ,MA,S PH,S	–	–	–	–	2
	2002	–	–	–	AZ,S (n=3)	AZ,S MA,S	–	–	–	5
	2003	C,DZ,S	–	–	AZ,S (n=6)	–	AZ,S	–	–	8
	2004	–	–	–	AZ,MA,S	–	–	–	–	1
	2005	C,S	–	–	AZ,S (n=2)	–	–	–	–	3
	2007	–	–	AZ,S (n=2)	–	–	–	–	–	2
	2009	DR,HX DR,S	DR,HX DR,HX,PY	CB,DR,HX CB,DR,HX,PZ DR,HX	CB,DR CB,DR,S DR,S	–	–	–	DR,PZ,S	11
Neal, upper, below diversion	2009	–	FL,HX PX,HX,MM	DR,HX	–	—	–	–	–	3
Rogers	2009	–	PX,EN,MM	–	–	–	–	–	–	1
Total mixture samples		22	17	17	36	4	3	10	7	116

Table 10. Minimum, median, and maximum concentrations of pesticides detected in samples from the 2004–05 Hood River basin, Oregon, fruit packers' study and number of samples with pesticides detected or not detected.

[**Abbreviations:** µg/L, microgram per liter; –, pesticide not detected or too few detections to calculate the median]

Sample type	Receiving stream	Pesticide	Detected concentration (µg/L)			Median detected concentration at mouth of receiving stream[1]	Detections	Nondetections
			Minimum	Median	Maximum			
Effluent	Lenz	Azinphos-methyl	0.05	0.95	37	0.12	11	5
		Phosmet	0.046	0.097	0.25	0.059	6	10
		Malathion	–	–	–	–	0	16
		Simazine	0.028	0.041	0.053	0.045	3	13
	Neal	Azinphos-methyl	–	–	–	0.059	0	1
		Phosmet	–	–	–	–	0	1
		Malathion	–	–	–	–	0	1
		Simazine	–	–	–	0.065	0	1
	Odell	Azinphos-methyl	18	–	18	no data	1	1
		Phosmet	0.65	–	0.65	no data	1	1
		Malathion	–	–	–	no data	0	2
		Simazine	–	–	–	no data	0	2
	Unnamed	Azinphos-methyl	4.7	–	12	no data	2	1
		Phosmet	0.85	–	17	no data	2	1
		Malathion	–	–	–	no data	0	3
		Simazine	–	–	–	no data	0	3
Surface Water	Hood	Azinphos-methyl	0.03	–	–	–	1	1
		Phosmet	–	–	–	–	0	2
		Malathion	–	–	–	–	0	2
		Simazine	–	–	–	–	0	2
	Lenz	Azinphos-methyl	0.14	1.3	4.8	0.12	9	1
		Phosmet	0.063	0.14	1.8	0.059	4	6
		Malathion	0.041	–	0.041	–	1	9
		Simazine	0.029	0.062	0.17	0.045	6	4
	Neal	Azinphos-methyl	0.063	–	0.063	0.059	1	4
		Phosmet	0.025	–	0.025	–	2	3
		Malathion	0.015	–	0.015	–	2	3
		Simazine	0.028	0.04	0.42	0.065	3	2
	Odell	Azinphos-methyl	0.033	0.045	0.26	no data	5	5
		Phosmet	0.025	0.043	0.043	no data	3	7
		Malathion	–	–	–	no data	0	10
		Simazine	0.024	0.027	2.4	no data	5	5
	Unnamed	Azinphos-methyl	–	–	–	no data	0	1
		Phosmet	–	–	–	no data	0	1
		Malathion	–	–	–	no data	1	0
		Simazine	0.068	–	0.068	no data	1	0

[1]Ambient concentrations are from June and September 2004 and September and October 2005, the months when the fruit packing effluent samples were collected.

Table 11. Detections in samples from the Hood River basin, Oregon, fruit packers' study exceeding U.S. Environmental Protection Agency and Oregon freshwater aquatic life standards.

[**Abbreviations:** USEPA, U.S. Environmental Protection Agency; CMC, criteria maximum concentration; CCC, criterion continuous concentration; –, no water-quality standard]

| | USEPA Office of Pesticide Programs Aquatic Life Benchmarks | | | | USEPA Water Quality Criteria | | Oregon Water Quality Criteria | | Percentage exceeding lowest standard (%) |
| | Fish | | Invertebrates | | | | | | |
	Acute	Chronic	Acute	Chronic	CMC (Acute)	CCC (Chronic)	Acute	Chronic	
Azinphos-methyl	0.18	0.055	0.08	0.036	–	–	–	0.01	60
Effluent									34
Surface water									57
Phosmet		3.2	1.0	0.8	–	–	–	–	6
Effluent									9
Surface water									4
Malathion	0.295	0.014	0.005	0.000026	–	0.1	–	0.1	2
Effluent									0
Surface water									11
Simazine	3,200	960	500	2,000	–	–	–	–	0

Table 12. Minimum, median, and maximum concentrations of priority trace elements detected in Hood River basin, Oregon, surface waters and national and State water-quality standards and exposure responses from literature.

[**Source:** Oregon Department of Environmental Quality, 2004; U.S. Environmental Protection Agency, 2005a, 2009b. All concentrations reported in micrograms per liter. **Abbreviations:** USEPA, U.S. Environmental Protection Agency; CMC, criteria maximum concentration; CCC, criterion continuous concentration; wk, week; m, month; h, hour; d, day; LC50, 50 percent lethal concentration; —, not detected, no water-quality standard or no literature values]

Trace element	Description	Number of samples	Number of detections	Detected concentration			USEPA		Oregon		Literature values			
				Minimum	Median	Maximum	CMC (acute)	CCC (chronic)	Acute	Chronic	Concentration	Exposure duration	Response	Source
Aluminum	Dissolved	42	25	13	44	530	750	87	—	—	539	2 wk	Olfactory stimulation	Tierney and others, 2010
	Total	4	4	127	543.5	1,590	—	—	—	—	—	—	—	—
	Total recoverable	158	146	12	410.5	9,690	—	—	—	—	—	—	—	—
Cadmium	Dissolved	25	11	0.1	0.25	0.42	[1]0.32–1.23	[1]0.07–15	—	—	100	10 m	Olfactory stimulation	Tierney and others, 2010
											1.5–37.9	—	LC50	Buhl and others, 1991
	Total	4	1	0.11	0.11	0.11	—	—	—	—	—	—	—	—
	Total recoverable	204	32	0.1	0.145	0.43	—	—	[3]0.46–2.20	[3]0.26–0.76	—	—	—	—
Copper	Dissolved	25	3	0.83	1.1	1.49	[1,2]2.34–8.65	[1,2]1.77–30	—	—	1–200	30 m–4 h	Olfactory stimulation	Tierney and others 2010
											0.1–6.4	—	Avoidance	
											330	—	Attraction	
											2–20	3 h	Alarm response	
											0.18–2.1	3 h	Alarm response	Hecht and others, 2007
											0.75–2.4	20 m–21 d	Avoidance	
											5–25	6 d - indefinite	Migration	
											9–57	96 h	LC50	
	Total	4	4	0.77	2.72	3.54	—	—	[3]2.97–11	—	—	—	—	—
	Total recoverable	205	116	0.31	1.29	26	—	—	—	[3]2.34–7.64	—	—	—	—
Nickel	Dissolved	25	10	0.26	1.875	4.32	[1]94–304	[1]10–32	—	—	6	—	Attraction	Tierney and others, 2010
	Total	4	4	0.43	0.565	0.96	—	—	—	—	23.9	—	Avoidance	Tierney and others, 2010
	Total recoverable	156	103	0.2	0.65	3.7	—	—	[3]285–921	[3]32–102	—	—	—	—

Table 12. Minimum, median, and maximum concentrations of priority trace elements detected in Hood River basin, Oregon, surface waters and national and State water-quality standards and exposure responses from literature.—Continued

[**Source:** Oregon Department of Environmental Quality, 2004; U.S. Environmental Protection Agency, 2005a, 2009b. All concentrations reported in micrograms per liter. **Abbreviations:** USEPA, U.S. Environmental Protection Agency; CMC, criteria maximum concentration; CCC, criterion continuous concentration; wk, week; m, month; h, hour; d, day; LC50, 50 percent lethal concentration; —, not detected, no water-quality standard or no literature values]

Trace element	Description	Number of samples	Number of detections	Detected concentration			USEPA		Oregon		Literature values			
				Minimum	Median	Maximum	CMC (acute)	CCC (chronic)	Acute	Chronic	Concentration	Exposure duration	Response	Source
Silver	Dissolved	25	0	—	—	—	[1]0.12–1.34	—	—	—	11.1–19.2	—	LC_{50}	Buhl and others, 1991
	Total	4	3	0.21	0.3	0.3	—	—	—	—	—	—	—	—
	Total recoverable	156	10	0.23	0.27	0.43	—	—	[3]0.16–1.69	—	—	—	—	—
Zinc	Dissolved	24	10	0.78	51.95	278	[1]23–76	[1]24–33	—	—	5.6	—	Avoidance	Tierney and others, 2010
	Total	4	4	2.49	5.085	10.4	—	—	—	—	—	—	—	—
	Total recoverable	205	115	0.54	3.87	36.1	—	—	[3]23–76	[3]21–69	—	—	—	—

[1]Hardness-dependent criterion based on hardness range of 15-60 mg/L, the general range of hardness values available in the basin from 1999 through 2009. Criteria were calculated using USEPA (2005a) National Recommended Water-quality criteria Appendix B: Parameters for Calculating Freshwater Dissolved Metals Criteria That Are Hardness-Dependent. The USEPA reports these criteria based on 100 mg/L hardness as $CaCO_3$.

[2]USEPA criteria for copper were calculated using the hardness-dependent formula. The USEPA now uses the Biotic Ligand Model (BLM), which uses 11 parameters to calculate copper criteria. The BLM could not be used here because data for all 11 parameters in the basin do not exist.

[3]Oregon hardness-dependent criteria are based on older USEPA guidelines, which use different calculation parameters than USEPA currently uses. Oregon criteria refer to total recoverable metal concentration, while the USEPA uses dissolved concentration (Oregon Department of Environmental Quality, 2006).

Table 13. Minimum, median, and maximum concentrations detected in Hood River basin, Oregon, surface waters and national and State water-quality criteria for nonpriority trace elements.

[**Source:** Oregon Department of Environmental Quality, 2004; U.S. Environmental Protection Agency, 2005a, 2009b. All concentrations reported in micrograms per liter. **Abbreviations:** USEPA, U.S. Environmental Protection Agency; CMC, criteria maximum concentration; CCC, criterion continuous concentration; –, no water-quality standard or no literature values]

| Trace element | Description | Samples | Detections | Detected concentration | | | USEPA | | Oregon | | Literature values | | | |
				Minimum	Median	Maximum	CMC (acute)	CCC (chronic)	Acute	Chronic	Concentration	Exposure duration	Response	Source
Antimony	Total Recoverable	156	7	3.7	4	5.6	–	–	–	–	–	–	–	–
Arsenic	Total Recoverable	156	6	2	2.35	3	–	–	–	–	–	–	–	–
Barium	Dissolved	25	16	6.32	10.9	19.4	–	–	–	–	–	–	–	–
	Total	4	4	1.27	5.05	14.1	–	–	–	–	–	–	–	–
	Total Recoverable	156	156	0.53	4.23	25.7	–	–	–	–	–	–	–	–
Beryllium	Total	4	4	0.015	0.0315	0.041	–	–	–	–	–	–	–	–
	Total Recoverable	156	44	0.01	0.0205	0.337	–	–	–	–	–	–	–	–
Calcium	Calcium	1	1	6,410	6,410	6,410	–	–	–	–	–	–	–	–
	Dissolved	42	26	7,140	9,780	15,000	–	–	–	–	–	–	–	–
	Total	46	30	2,240	9,600	15,000	–	–	–	–	–	–	–	–
	Total Recoverable	207	204	2,150	6,685	16,600	–	–	–	–	–	–	–	–
Chromium (III)	Dissolved	25	13	0.2	0.76	1.28	[1]120–375	[1]16–25	–	–	–	–	–	–
	Total	4	2	0.28	0.605	0.93	–	–	–	–	–	–	–	–
	Total Recoverable	205	110	0.21	0.45	4.2	–	–	[2]367–1,143	[2]44–136	–	–	–	–
Cobalt	Dissolved	25	6	0.34	0.94	1.14	–	–	–	–	–	–	–	–
	Total	4	3	0.25	0.31	0.31	–	–	–	–	–	–	–	–
	Total Recoverable	156	52	0.2	0.32	1.52	–	–	–	–	24–180	–	Avoidance	Tierney and others, 2010
Iron	Dissolved	42	26	7.6	81.75	229	–	–	–	–	–	–	–	–
	Total	4	4	57.2	418	766	–	1,000	–	–	1,200	–	–	–
	Total Recoverable	407	204	12.9	358	6,220	–	–	–	1,000	4,250–6,450	–	Avoidance	Tierney and others, 2010
Lanthanum	Dissolved	42	1	1.1	1.1	1.1	–	–	–	–	–	–	–	–
	Total Recoverable	53	4	1.7	2.6	4.9	–	–	–	–	–	–	–	–
Lithium	Dissolved	42	26	1.15	5.795	12	–	–	–	–	–	–	–	–
	Total	4	4	1.64	3.97	6.35	–	–	–	–	–	–	–	–
	Total Recoverable	158	123	0.42	3.82	9.28	–	–	–	–	–	–	–	–
Magnesium	Unspecified	1	1	2,570	2,570	2,570	–	–	–	–	–	–	–	–
	Dissolved	42	26	3,020	6,050	8,850	–	–	–	–	–	–	–	–
	Total	4	4	738	1,310	4,070	–	–	–	–	–	–	–	–
	Total Recoverable	207	204	684	3,280	10,700	–	–	–	–	–	–	–	–

Table 13. Minimum, median, and maximum concentrations detected in Hood River basin, Oregon, surface waters and national and State water-quality criteria for nonpriority trace elements.—Continued

[**Source:** Oregon Department of Environmental Quality, 2004; U.S. Environmental Protection Agency, 2005a, 2009b. All concentrations reported in micrograms per liter. **Abbreviations:** USEPA, U.S. Environmental Protection Agency; CMC, criteria maximum concentration; CCC, criterion continuous concentration; –, no water-quality standard or no literature values]

| Trace element | Description | Samples | Detections | Detected concentration | | | USEPA | | Oregon | | Literature values | | | |
				Minimum	Median	Maximum	CMC (acute)	CCC (chronic)	Acute	Chronic	Concentration	Exposure duration	Response	Source
Manganese	Dissolved	42	26	1.67	20.95	67.8	–	–	–	–	–	–	–	–
	Total	4	4	1.73	10.59	46.8	–	–	–	–	–	–	–	–
	Total Recoverable	158	152	0.26	9.45	123	–	–	–	–	–	–	–	–
Molybdenum	Dissolved	25	3	0.5	0.5	0.61	–	–	–	–	–	–	–	–
	Total Recoverable	156	20	0.41	0.49	0.82	–	–	–	–	–	–	–	–
Potassium	Dissolved	42	26	1,100	1,660	2,640	–	–	–	–	–	–	–	–
	Total	4	4	412	855	1,510	–	–	–	–	–	–	–	–
	Total Recoverable	158	155	334	1,020	2,270	–	–	–	–	–	–	–	–
Selenium	Total Recoverable	156	21	3	3.9	6.7	–	5	260	35	–	–	–	–
Sodium	Dissolved	42	26	5,170	10,520	136,000	–	–	–	–	–	–	–	–
	Total	4	4	1,690	2,935	5,340	–	–	–	–	–	–	–	–
	Total Recoverable	158	155	1,550	4,380	25,200	–	–	–	–	–	–	–	–
Thallium	Dissolved	25	6	3.3	6.05	9.3	–	–	–	–	–	–	–	–
	Total	4	2	5.7	–	6.4	–	–	–	–	–	–	–	–
	Total Recoverable	156	8	2	2.6	7	–	–	–	–	–	–	–	–
Vanadium	Dissolved	26	16	3.01	7.07	8.2	–	–	–	–	–	–	–	–
	Total	4	4	1.01	2.54	4.7	–	–	–	–	–	–	–	–
	Total Recoverable	156	155	0.61	2.87	9.08	–	–	–	–	–	–	–	–

[1]Hardness-dependent criterion based on hardness range of 15-60 mg/L, the general range of hardness values available in the basin from 1999 through 2009. Criteria were calculated using USEPA (2005a) National Recommended Water-quality criteria Appendix B: Parameters for Calculating Freshwater Dissolved Metals Criteria That Are Hardness-Dependent. The USEPA reports these criteria based on 100 mg/L hardness as $CaCO_3$.

[2]Oregon hardness-dependent criteria are based on older USEPA guidelines, which use different calculation parameters than USEPA currently uses. Oregon criteria refer to total recoverable metal concentration, while the USEPA uses dissolved concentration (Oregon Department of Environmental Quality, 2006)

Discussion

When the ODEQ began pesticide monitoring in 1999, the organophosphate insecticides azinphos-methyl and chlorpyrifos were detected in several creeks at concentrations exceeding the State of Oregon's chronic or acute criteria for the protection of freshwater aquatic life. In 2001, additional sites were added to the stream sampling network and more samples were collected at each site to better characterize the sources and temporal character of organophosphate transport (appendix B). Stream monitoring has continued through 2010. Throughout the monitoring period, sites were dropped and new ones added in an attempt to characterize pesticide occurrence and distribution in the major salmon-bearing streams in the Hood River basin (appendix H).

Coincident with the monitoring program, the ODEQ, Oregon Department of Agriculture, Hood River Soil and Water Conservation District, the Confederated Tribes of Warm Springs, and the Oregon State University agricultural extension service began an outreach and education program to work with farmers to reduce pesticide drift and runoff from land in their stewardship. The early to mid-2000s also marked the beginning of a period of use restrictions and cancellations for many organophosphates, including azinphos-methyl and chlorpyrifos, as the USEPA implemented the Food Quality Protection Act of 1996 and as less-toxic alternatives to organophosphates became available (Grafton-Cardwell and others, 2005; U.S. Environmental Protection Agency, 2006a, 2006c, 2009c). Pesticide use was further restricted in Oregon, Washington, and California in 2004 by a ruling by the U.S. District Court for the Western District of Washington in the case of Washington Toxics Coalition v. EPA (U.S. Environmental Protection Agency, 2004), which restricted the application of 26 pesticides adjacent to streams used by salmon that are listed as threatened or endangered under the Endangered Species Act. Azinphos-methyl, chlorpyrifos, and 10 other organophosphates were among the pesticides affected by this ruling.

Pesticides Detected Since 2007

Since sampling commenced in 1999, the frequency of detecting most pesticides that were monitored for the entire period appears to have declined. An exact measure of the decline is unknown due to changes in reporting limits, sites, the number of samples collected each year, and the time when samples were collected during the year. In the unscreened dataset, neither diazinon nor malathion has been detected since 2005. Phosmet was detected twice from 2006 through 2009 compared with 11 detections from 2002 through 2005. Chlorpyrifos was detected eight times from 2006 through 2009; however, seven of these detections occurred during a 2-week period in April 2008, and there were no detections in 2009. Preliminary ODEQ data from 2010 monitoring show chlorpyrifos detections in four creeks in March 2010, once exceeding the lowest national and State water-quality criteria (Kevin Masterson, Oregon Department of Environmental Quality, written commun., 2010). Azinphos-methyl and simazine continue to be detected, but are primarily limited to Lenz Creek at mouth and Neal Creek at mouth.

Azinphos-methyl

Sixty-nine of the 76 detections of azinphos-methyl in the unscreened dataset occurred at Lenz Creek at mouth and Neal Creek at mouth. Discussion will be limited to these two sites. Azinphos-methyl primarily was detected in Lenz Creek and Neal Creek between late-May and mid-June and from mid-August through at least mid-October. No samples were collected after mid-October, so the occurrence of azinphos-methyl in the streams of Hood River basin is unknown from mid-October until March. Detections in May and June correspond to the primary period of use of the insecticide in the Hood River basin (Eugene Foster, Portland State University, written commun., 2003; Jenkins, 2003). Trends in detected azinphos-methyl and chlorpyrifos concentrations in May and June of 2000 and 2001 reported here were consistent with the findings of another study conducted in the same area during the same period (Jenkins, 2003). The presence of azinphos-methyl in the creeks during this period is likely the result of spray drift and runoff of irrigation water from treated fields, although neither has been directly measured. Jenkins (2003), who hypothesized that runoff following precipitation events is a major contributor of pesticides to streams, measured instream concentrations of azinphos-methyl and chlorpyrifos and did not find a significant relationship with precipitation. However, use of those pesticides on adjacent agricultural land was not measured. Late-summer concentrations of azinphos-methyl in Lenz Creek tended to be higher than concentrations in May and June; concentrations were similar during both periods in Neal Creek. The higher concentrations in August, September, and October in Lenz Creek likely were due to the discharge of wash water by fruit washing and packing facilities that line that creek (table 10). The median concentration of azinphos-methyl in samples of wastewater discharged from these facilities into Lenz Creek in 2004 and 2005 was 0.95 µg/L; concentrations as high as 37 µg/L were measured in wastewater. Lower instream discharge (flow) rates during those months could also contribute to the higher detected concentrations. Although discharge data are not available for Lenz Creek, other streams in the basin have the lowest mean monthly discharge during July to October (U.S. Geological Survey, 2010).

Azinphos-methyl was detected in 66 samples from Lenz Creek at mouth and Neal Creek at mouth from 1999 through 2006. From 2007 through 2009, there were 3 detections.

The dramatic decrease in the occurrence of azinphos-methyl cannot, with certainty, be ascribed to decreasing instream concentrations. Azinphos-methyl was commonly detected at these sites in June, during the peak application period; however, just one sample was collected in June 2007 and no samples were collected in June 2008. Weekly samples were collected at the two sites in June 2009. Similarly, a shift in the time of sample collection after 2006 resulted in just five samples (two at Lenz Creek at mouth, three at Neal Creek at mouth) being collected during the months of the historically highest instream azinphos-methyl concentrations—August, September, and October.

Azinphos-methyl was not detected at Lenz Creek or Neal Creek in June 2009. The reporting limit for azinphos-methyl was stable during this period and lower than it was in the early to mid-2000s. These data suggest that the causes of azinphos-methyl transport to the creeks during the June application period have been addressed, but sampling frequency was not ideal to make this assertion with certainty. There was one detection of azinphos-methyl in September 2009 in Lenz Creek, suggesting that there may still be reason for concern about effluent from the fruit washing and packing houses. Though rarely detected, concentrations of azinphos-methyl detected since 2007 have all exceeded the chronic Oregon water-quality criteria. Azinphos-methyl is set to be phased out on its last registered uses (alkali bee beds, apples, blueberries, cherries, parsley, and pears) by September 30, 2012 (U.S. Environmental Protection Agency, 2009c).

Simazine

Simazine was detected in the unscreened 2007–2009 data at about the same frequency and concentration at which it was detected in 2005 and 2006. Most detections were at Lenz Creek at mouth and Neal Creek at mouth. Concentrations of simazine detected since 2007 range from 0.0063 to 0.299 µg/L and are within an order of magnitude of those found to cause a reduced olfactory response to a female priming pheromone in adult male Atlantic salmon (Moore and Lower, 2001). Two recent studies using common carp (*Cyprinus carpio*) have demonstrated histopathological changes in liver and kidney tissues in fish exposed to simazine concentrations as low as 4 µg/L (Velisek and others, 2009) and 42 µg/L (Oropesa and others, 2009). The effect on salmonids has not been documented. Other studies have demonstrated that high concentrations of simazine (greater than 25 µg/L) increase the toxicity of organophosphates to aquatic invertebrates (Schuler and others, 2005; Trimble and Lydy, 2006). The highest simazine concentration measured in any Hood River stream was 1.9 µg/L at Lenz Creek at mouth in 2003. The highest concentration observed in 2009 was 0.299 µg/L, also at Lenz Creek at mouth. These concentrations are within an order of magnitude of documented deleterious effects in fish and aquatic invertebrates. While detected concentrations of

simazine are not necessarily cause for concern on their own, simazine use in the basin is common, and it has the potential to compound the toxicity of more deleterious pesticides.

New Pesticides for 2009

In 2009, 14 pesticides were detected in the 8 streams that were sampled. Of these 14 pesticides, 12 were new pesticides added to the suite of analyses in 2009. Pesticides were detected in at least one sample collected from every site except West Fork Hood River at Moving Falls (RM 2.5). Data from 2010 analyzing the same suite of 100 pesticides became available in early 2011, but their analysis is outside the scope of this study. The data are available through the ODEQ's LASAR database (Oregon Department of Environmental Quality, 2008).

Eight of the new pesticides detected in 2009 are of little concern due to (1) their infrequent or irregular occurrence (table 7) and (2) their low concentrations relative to established water-quality standards and potential for deleterious sublethal effects on aquatic organisms (figs.11–22). With the exception of the single detection of endrin at Rogers Spring Creek, the maximum concentrations of the 12 newly identified pesticides were at least one order of magnitude lower than established water-quality standards and concentrations that can cause sublethal toxicity to salmonids and aquatic invertebrates. However, the effects of these pesticides at the measured concentrations are unknown when they occur in mixtures. Additional discussion of four pesticides—carbaryl, diuron, endrin, and hexazinone—is warranted due to the relatively high frequency of detection or toxicity.

Carbaryl

Carbaryl was primarily detected in May and June. Seven of the nine detections occurred at Lenz Creek at mouth or Neal Creek at mouth. Although the maximum concentration of carbaryl was 14 times lower than the chronic USEPA criterion for the protection of freshwater invertebrates, the pesticide has been shown to additively increase acetylcholinesterase inhibition in the presence of other carbamates and organophosphate insecticides at low concentrations (Laetz and others, 2009). Acetylcholinesterase (AChE) inhibition is a commonly used biomarker of exposure to organophosphate and carbamate pesticide exposure in fish (Eugene Foster, Portland State University, written commun., 2003; Scholz and others, 2006). Inhibition of the AChE enzyme has been associated with impairment of swimming, predator detection and avoidance, and migration (Jarrard and others, 2004; Sandahl and others, 2005). The low concentrations of carbaryl at these two sites may not be of concern if concentrations of organophosphates in May and June remain less than aquatic life criteria, as they were in 2009.

Diuron

Diuron was detected at all sites except West Fork Hood River at Moving Falls (RM 2.5), at which only a single sample was collected. It was detected throughout the year at Lenz Creek at mouth, Hood River at mouth, Neal Creek at mouth, and Middle Neal Creek at Hwy 35. Concentrations of diuron were always greater than an order of magnitude (10 times) less than the USEPA chronic criteria for the protection of freshwater fish; most concentrations were more than a factor of 100 lower. Thus, the presence of diuron in the sampled streams is unlikely to pose a threat to aquatic organisms unless it occurs in mixtures of pesticides that potentiate the effects of one or more of those pesticides. 3,4-dichloroaniline (DCA) is a degradation product of diuron that was not analyzed for this project. The acute and chronic toxicities of DCA are lower than those for diuron (Crossland, 1990; U.S. Environmental Protection Agency, 2003). Sublethal effects have been observed in fish at concentrations of about 200 µg/L (Crossland, 1990; Munn and others, 2006).

DCA detections have been common in Oregon streams (U.S. Geological Survey, 2010). In 460 surface-water samples collected in Oregon since 1990, DCA was detected in 150. In samples where both DCA and diuron were detected, the concentration of diuron exceeded the concentration of DCA by a factor of 7 to 39, and DCA was never detected without diuron also being detected. Assuming that DCA is present in all samples collected from the Hood River basin in 2009 that contained diuron, and assuming that the ratio of DCA to diuron in Hood River streams is similar to that observed in other streams in Oregon, it is reasonable to conclude that mortality and sublethal effects resulting from the presence of DCA is probably not a concern in the streams sampled in 2009.

Endrin

The single detection of endrin in 2009 at Rogers Spring Creek is unusual, and serves as a reminder of the persistence of many older pesticides in the environment. The USEPA cancelled the last registered agricultural use of endrin in 1985 (U.S. Environmental Protection Agency, 2000); however, the half-life of endrin in soil can be up 14 years (Agency for Toxic Substances and Disease Registry, 1996). Endrin strongly binds to soil and is unlikely to be found in the dissolved phase; the adsorption coefficient (K_{oc}) is approximately 34,000 (Agency for Toxic Substances and Disease Registry, 1996). The single detection could have been related to runoff from rainfall that occurred on the day prior to sample collection; however, endrin was not detected in other samples collected after other rain events. Alternately, construction, planting, or tilling could have disturbed soil contaminated with endrin, which was then washed into the creek.

Hexazinone

With one exception, all detections of hexazinone occurred at three sites along Neal Creek. Concentrations at Upper Neal Creek below agricultural diversion and Middle Neal Creek were similar for most of the year, and with one exception, were always greater than the concentration at Neal Creek at mouth (fig. 32). The decrease in concentration between Middle Neal Creek and Neal Creek at mouth was probably due to dilution by tributary inflows (such as Lenz Creek) and possibly upwelling groundwater along this reach of the creek. A more detailed examination of the drop in concentration is limited by an absence of flow data at these sites.

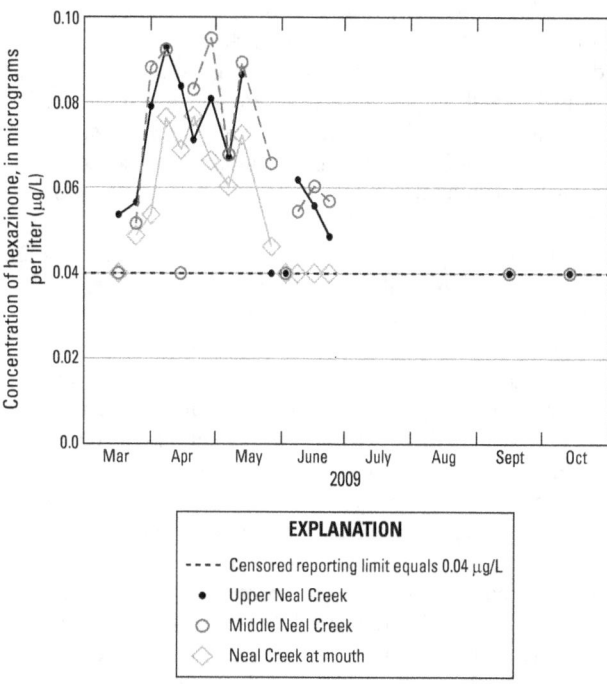

Figure 32. Hexazinone concentrations at three Neal Creek, Oregon, sites in 2009.

In the Hood River basin, most, if not all, hexazinone is used on forest land. It is registered for use along rights-of-ways, but there is no record of its use for this purpose in the Hood River basin (Brian Walker, Oregon Department of Transportation, oral commun., 2010; John Buckley, East Fork Irrigation District, oral commun., 2010; Nate Lain, Hood River County Weed and Pest Division, oral commun., 2010). It also can be applied to alfalfa, grass hay, noncrop agricultural areas, and industrial areas, but these represent small areas of the Hood River basin. The land upstream of Upper Neal Creek is 95 percent forest and the land draining to Neal Creek between the Upper Neal Creek site and the Middle Neal Creek site is 66 percent forest and 37 percent agricultural (appendix A). Considering the limited major users of hexazinone and the land use contributing to the sampling sites where it was detected, forestry use is the most probable source of the hexazinone in Neal Creek. Preliminary 2010 data showing frequent detections of imazapyr, another forestry herbicide, at the three Neal Creek sites and only one detection of hexazinone (Kevin Masterson, Oregon Department of Environmental Quality, written commun., 2010) reflect the annual variability in forestry herbicide use in response to changing needs.

Concentrations of hexazinone detected in Neal Creek are 5 to 6 orders of magnitude less than established water-quality benchmarks. Few studies using environmentally relevant concentrations of hexazinone (< 1 mg/L) exist in the literature. Nieves-Puigdoller and others (2007) found concentrations of hexazinone 100 µg/L had no effect on smolt development in Atlantic salmon. Michael and others (1999) found no change in the aquatic invertebrate community after hexazinone application in a forested watershed. Concentrations as high as 473 µg/L were observed. In lab studies using mammalian test subjects, developmental and reproductive toxicity were observed only at concentrations approaching the 50 percent lethal dose (U.S. Environmental Protection Agency, 1994).

Considering the available evidence, hexazinone at concentrations observed in streams of Hood River basin in 2009 is probably not a concern. However, the pesticide was present in prime salmon-rearing habitat during at least 4 months in 2009; additional research to more confidently determine the sublethal effects of hexazinone on Pacific Northwest salmonids using environmentally relevant concentrations would aid in the assessment of risk to these species.

Pesticide Mixtures in 2009

Thirty-four of the 111 samples collected in 2009 contained at least 2 pesticides (table 9). Mixtures of 2 pesticides were the most common (20 samples) followed by mixtures of 3 pesticides (10 samples). Mixtures of more than 3 pesticides were identified in 4 samples. Three herbicides were among the most common pesticides identified in mixture samples: diuron was a component of 28 mixture samples, hexazinone was a component of 15 mixture samples, and simazine was a component of 13 mixture samples. The insecticide carbaryl was a component of 8 mixture samples. The most common mixture in 2009 was of the herbicides diuron and simazine, which were found together in 12 samples.

Mixtures of pesticides are a concern because of the unknown manner in which the chemicals can affect an organism. Dose-addition models are commonly used to predict cumulative toxicity for co-occurring pesticides. These models predict the toxicity of a mixture by adding the toxic potency of each component in the mixture. Some pesticides have been shown to interact synergistically, resulting in greater toxicity than predicted by a simple dose-addition model (Lydy and Austin, 2004; Schuler and others, 2005; Trimble and Lydy, 2006; Laetz and others, 2009). Conversely, some pesticide mixtures may result in lower toxicity than predicted by a dose-addition model (Key and others, 2007; Brander and others, 2009).

The effects of pesticide mixtures are an area of active research among ecotoxicologists, and the USEPA is developing pesticide regulations that address mixtures of pesticides that have a common mode of action. However, the vast number of possible mixtures of chemicals in the environment combined with differing modes of action makes the issue of cumulative pesticide toxicity particularly difficult to address. Among the better studied mixtures are those involving one or more organophosphate insecticides. Synergistic toxicity to salmonids has been shown for mixtures of carbamate and organophosphate insecticides (National Marine Fisheries Service, 2008). Some triazine herbicides have been shown to potentiate the toxicity effects of organophosphate insecticides on aquatic invertebrates that are salmonid prey items, and thus could indirectly affect salmonids (Lydy and Austin, 2004; Trimble and Lydy, 2006). More work is needed to understand the effects of the mixtures of pesticides commonly observed in streams of the Hood River basin.

Trace Elements

Samples for trace elements were collected and analyzed at 53 sites. Most sites were sampled once or twice; most data were collected from 1999 through 2002. The data provide a preliminary screen of potential salmonid toxicity related to trace elements, but a comprehensive analysis of the data are limited by several factors: (1) most sites were sampled only once or twice, (2) samples were not collected at the same time of year at all sites, (3) samples were not collected throughout the year, (4) the data may not reflect current conditions, (5) many toxicity criteria are dependent upon the hardness of the water, which is not available for these samples, and (6) most samples were analyzed for total recoverable trace elements rather than the dissolved fraction.

With the caveats just noted, it is possible to develop some working hypotheses. A preliminary literature review indicated that concentrations are likely not of concern for most of the trace elements analyzed. However, the median concentrations of the following trace elements exceeded or were within an order of magnitude of criteria established by USEPA or the State of Oregon: aluminum (dissolved), cadmium (dissolved and total recoverable), copper (dissolved and total recoverable), iron (total recoverable), nickel (dissolved), selenium (total recoverable), silver (total recoverable), and zinc (dissolved and total recoverable) (tables 12 and 13). The review identified studies that have documented lethal and sublethal effects of trace elements on salmonids and aquatic invertebrates, at concentrations similar to and higher than State and national water-quality standards. Dissolved concentrations of aluminum, cadmium, copper, iron, nickel, and zinc were detected at concentrations exceeding one or more values in this review (tables 12 and 13). Dissolved data are only available for 26 sites and consist of one sample per site collected in October 1999.

This preliminary screen suggests that some trace elements might occur at concentrations of concern for salmonids or their prey. More work is needed to ascertain their duration and spatial extent in creeks of the Hood River basin and to determine sources and transport mechanisms.

Status of Prey Organisms

The ODEQ has conducted annual surveys of benthic invertebrates at five sites in the Hood River basin since 2000. In an analysis of these data through 2007, ODEQ researchers suggested that there is a spring depression in the invertebrate community following periods of pesticide application and that the community recovers later in the season (Shannon Hubler, Oregon Department of Environmental Quality, written commun., 2008). There is larger within-season variability in the observed-over-expected (O/E) scores than there is over the period of record, making it difficult to determine whether the communities have improved over time. Further, based on their data, it is not possible to understand differences in the communities among sites or changes in community structure over time; information on the type of prey insects and their abundance would be useful.

Information Gaps

The Hood River Watershed Group's Watershed Action Plan identifies projects and strategies "to improve watershed health, water quality, and fish populations in the Hood River Watershed" (Hood River Watershed Group, 2008). To achieve this goal, a comprehensive evaluation of the spatial and temporal occurrence of all potential toxic contaminants and their effects on the instream biota is needed. Data collected by the HRPSP since 1999 represent important surveys; however, significant gaps still exist.

Spatial Distribution of Contaminants

Most streams (or a nearby downstream reach) that have been identified as critical habitat for steelhead and salmon have been sampled for at least 1 year (appendix H). Several, however, have not, including Green Point Creek, Tony Creek, and Emil Creek. Although Odell Creek is not identified as critical habitat, steelhead have been observed recently in that stream (Joe McCanna, the Confederated Tribes of Warm Springs, oral commun., 2010); the ODEQ began sampling this creek in 2010.

The HRPSP monitoring sites have changed over time. Seven of 16 sites have not been monitored for pesticides since 2006 or earlier (appendix H). Some sites were dropped because of redundancy or because better sites were identified; for example, the two sites on upper Neal Creek. However, some sites were dropped because pesticides were infrequently detected or concentrations were consistently near or less than the reporting limit, such as at Baldwin Creek. Budgetary constraints also limited the number of sites sampled each year. The current network of sampling sites does not comprehensively cover all critical habitat streams (StreamNet, 2010).

The number of sites at which trace elements were sampled and analyzed was larger than the number for pesticides. Most streams identified as critical habitat for steelhead and salmon were sampled. Only five samples for trace elements have been collected from those streams since 2002.

Seasonal Distribution of Contaminants

The ODEQ pesticide monitoring during the past decade was scheduled to coincide with peak pesticide use in the basin (March–June and September). Ninety percent of samples were collected during these 5 months (appendix B). However, pesticides also were detected in February, July, August, October, and December. No pesticides have been detected in November, but only five samples were collected during this month among all sites since 1999. No samples have been collected in January. Pesticides are potentially used throughout much of the year in the basin's major land uses (appendix I), and several salmonid species (winter and summer steelhead, spring Chinook, coho, and bull trout) are present in streams year-round (National Marine Fisheries Service, 2008).

Most sites were sampled only once (in October) for trace elements. Ten other sites were sampled mainly in March–June. It is reasonable to expect seasonal differences in the concentration of some trace elements. For example, concentrations of trace elements associated with automobiles, such as cadmium, copper, cobalt, iron, nickel, lead, and zinc have been observed to increase seasonally in response to increased runoff from roadways (Hallberg and others, 2007).

Pesticides Used in the Hood River Basin

Many pesticides are commonly used in the Hood River basin or are registered for use for the basin's major land uses, but have not been analyzed for in its surface waters (appendix C and appendix J). Neonicotinoids are a class of insecticides that were developed as replacements for organophosphate, carbamate, and synthetic pyrethroid insecticides. Use of neonicotinoids and other organophosphate replacements has increased in the basin (Steve Castagnoli, Oregon State University Extension Service, oral commun., 2010), yet many are not included in the 2009 suite of analyzed pesticides. Likewise, many pyrethroid insecticides—a class that is in common use in the basin—have not been analyzed.

Hydrophobic Pesticides

Hydrophobic (particle-bound) pesticides bind strongly to sediments and plant matter and are less likely to be found dissolved in the water column. The presence of particle-bound pesticides in Hood River basin streams could be underrepresented in the current dataset, which only includes water samples, particularly in streams with fine-grained or organic-rich sediments. This is especially relevant for pyrethroid insecticides, which are used in the basin and bind more strongly to particles than most current-use pesticides. Pyrethroid insecticides have been associated with sediment toxicity to benthic invertebrates (Weston and others, 2004; Amweg and others, 2005, 2006; Holmes and others, 2008; Domagalski and others, 2010), are highly toxic to salmonids (Marking, 1974; Coats and O'Donnell-Jeffrey, 1979; Kumaraguru and Beamish 1981; Ural and Sağlam, 2005), and have been shown to interfere with reproductive behavior in brown trout (Jaensson and others, 2007).

Organochlorine pesticides are also hydrophobic. Most organochlorine pesticides, such as aldrin, chlordane, DDT, dieldrin, endrin, and lindane, have been banned in the United States due to their persistence and toxicity. However, they are often found in fish and sediment throughout the Columbia Basin. Two of these pesticides, DDT and lindane, were detected in Hood River and Neal Creek sediments in May 1998, when sediments from five sites were sampled for pesticides (Oregon Department of Environmental Quality, 2008). Endrin was identified in a surface-water sample from Rogers Spring Creek in 2009.

Pesticide Mixtures

Many samples contained at least two pesticides in the same sample. The number of pesticides actually present in any sample of water is potentially even larger because the instream presence of many pesticides used in the Hood River basin is unknown (appendix J). Moreover, instantaneous grab samples can fail to detect pesticides that are intermittently present in a stream; Jenkins (2003) found considerable variability in concentrations of organophosphates measured over periods of hours and days in Neal Creek. An assumption of simple dose-additivity provides a mechanism for initially assessing the potential toxic effects of mixtures. However, even a simple model such as this is limited by the lack of experimental research using pesticide mixtures at concentrations that are environmentally realistic for the Hood River basin.

Trace Elements

Trace-element concentration data are approximately 10 years old and exist only for limited parts of a few years (October 1999 and March–July 2000–01). The 1999 samples were analyzed for dissolved trace elements, whereas the 2000–01 samples were analyzed for total or total recoverable trace elements. Total and total recoverable analyses include dissolved constituents and trace elements contained in suspended particulate matter; the latter are less biologically available. USEPA water-quality criteria apply to dissolved trace-element concentrations, whereas Oregon criteria apply to total recoverable concentrations. Data from October 1999 indicate that concentrations of some dissolved trace elements with potential to harm salmonids approached or exceeded Oregon and/or USEPA criteria (dissolved aluminum in Wisehart Creek, cadmium in McGuire and Odell Creeks, and zinc in Lenz Creek). The concentration of some trace elements of concern also approached or exceeded values cited in toxicology literature that were shown to elicit olfactory stimulation or avoidance responses in salmonids (dissolved aluminum in Wisehart Creek and copper and zinc in Lenz Creek) (Tierney and others, 2010). However, the duration of the detected concentrations is unknown and could be less than the time periods upon which the criteria or toxicology studies were based.

Groundwater Contamination

Agricultural and urban use of pesticides has been correlated to their presence in groundwater. The presence of hydrophilic pesticides (those likely to be found dissolved in water) in surface waters of the Hood River basin indicates that there is potential for groundwater contamination. However, the presence and distribution of pesticides in Hood River groundwaters is unknown and outside the scope of this report. Nevertheless, discharge of contaminated groundwater can contribute to pesticide loading in streams (Ebbert and Embrey, 2002). Subsurface and surficial contributions of pesticides to streams were not assessed in this report and would require a sampling plan designed to address this question.

Unsampled Contaminants

Although concern about organophosphate insecticides is waning with their decreasing use, concern about the effects of degraded water quality on threatened salmonids in the basin remains. Numerous chemicals currently or historically used in the Hood River basin have not been analyzed in streams and bed sediment; however, nonpesticide chemicals are not within the scope of the HRPSP project. Throughout the lower Columbia River, polychlorinated biphenyl (PCB) compounds, polycyclic aromatic hydrocarbon (PAH) compounds, mercury, and legacy organochlorine pesticides (such as DDT) are commonly found in fish and birds in Western Oregon (Hinck and others, 2004; Johnson and others, 2007; Henny and others, 2008; Sherman and others, 2009). Additionally, laboratory analytical techniques developed in the last decade now enable scientists to examine contaminants such as pharmaceuticals, synthetic estrogens and androgens, and a variety of other potentially toxic compounds that enter streams from wastewater treatment plants, septic systems, and runoff from developed areas. Sampling conducted in the lower Columbia River basin in 2007 identified many of these chemicals in water, fish, and bed sediment (Lower Columbia River Estuary Partnership, 2007; Nilsen and others, 2007). The environmental fate and toxicity to aquatic biota for many of these novel analytes are still being determined. Analysis of samples from the Hood River wastewater treatment plant and samples of stormwater runoff in the city of Hood River collected in 2008–09 is under review by the U.S. Geological Survey. Those data could provide the first information on many of these compounds from a location in the Hood River basin. Results of additional sampling for PCB compounds, PAH compounds, polybrominated biphenyl ethers, organochlorine pesticides, and mercury in the water column and fish tissue during the summer of 2009 from 31 sites along the middle Columbia River and its major tributaries, including Hood River, are scheduled to be available from the ODEQ in 2011 (Kevin Masterson, Oregon Department of Environmental Quality, written commun., 2011).

Invertebrate Data

Collecting macroinvertebrate assemblages is a common, relatively inexpensive method for assessing biological integrity of streams. Foster and others analyzed macroinvertebrate data collected in 2002 in the Hood River basin (Eugene Foster, Portland State University, written commun., 2003). They found differences in macroinvertebrate assemblages among forested and agricultural sites. They also showed differences in assemblages at the same site before and after insecticide spraying in orchards within the catchments draining to the sampling sites. Macroinvertebrate data have been collected during the spring and summer from 2000 through 2008 at seven sites that were also monitored for pesticides. Although beyond the scope of this report, further analysis of these data could be useful in assessing trends in the invertebrate communities where best management practices have occurred.

Summary and Conclusions

Since 1999, the Hood River Pesticide Stewardship Partnership (HRPSP) has encouraged voluntary adoption of best management practices to reduce water-quality impacts associated with pesticide use in the Hood River basin. The Oregon Department of Environmental Quality collected and analyzed pesticide samples every year since the inception of the HRPSP to monitor the distribution and concentrations of pesticides in the basin's salmonid-bearing streams. Water sampling coincided with peak pesticide application in the basin; 90 percent of samples were collected in March through June and in September.

Seven of 10 pesticides analyzed from 1999 through 2009 were detected at least once: two triazine herbicides and five organophosphate insecticides. Most pesticide detections were at Lenz Creek at mouth and Neal Creek at mouth. Simazine was the herbicide detected most frequently and at the highest concentration. It was detected at approximately the same frequency since 2007 as prior to 2007. Simazine was present at concentrations within an order of magnitude (approximately a factor of 10) of those known to cause sublethal effects on fish and aquatic invertebrates. Azinphos-methyl was the most frequently detected insecticide and typically was measured at the highest concentrations, although the maximum detected insecticide concentration was for chlorpyrifos. Azinphos-methyl was detected at concentrations greater than the Oregon water-quality criterion for chronic exposures every year except 2008. The presence of azinphos-methyl in Lenz Creek in September 2009 indicates that effluent water from fruit-packing facilities remains a potential source of organophosphate insecticide contamination to streams.

The frequency of detection of pesticides monitored since 1999 has declined; however, the analysis of trends in detections was confounded by two main factors: (1) monitoring was not consistent across sites, years, or months, and (2) reporting limits changed within and across years. For example, reduced detection frequency could be caused by (a) true reductions of instream concentrations, (b) a decreased number of samples collected since 2007 during months when, in preceding years, pesticides were detected most frequently or were detected at the highest concentrations, or (c) increased reporting limits in later years.

Fourteen of 100 pesticides analyzed in 2009 were detected; 12 of those were analyzed for the first time in 2009. Eight of the detected pesticides were measured at low concentrations relative to those known to cause toxicity or sublethal effects to aquatic organisms or were detected too infrequently to warrant concern. Endrin was the only pesticide detected for the first time in 2009 at a concentration exceeding any national or State water-quality criterion.

Instream mixtures of pesticides can cause less-than-additive, additive, or synergistic (greater-than-additive) toxicity to aquatic organisms. Triazine herbicides and carbamate insecticides are classes known to potentiate organophosphate toxicity to aquatic invertebrates and salmonids, respectively. Thus, even at low concentrations, the presence of some detected pesticides is of concern because of their synergism with organophosphate pesticides. Simazine (triazine) and carbaryl (carbamate) were among the most common pesticides detected in mixture samples in this dataset. In 1999–2009, 12 percent of samples had two or more pesticides detected in the same sample. In 2009, 31 percent of samples had mixtures of pesticides detected. The increase in 2009 likely was due to the expanded list of pesticides analyzed that year. Both values likely under-represent the presence of pesticide mixtures in streams since many pesticides known to be used in the basin were not analyzed for this project. Effects to salmonids of the observed pesticide mixtures at concentrations detected in the basin are unknown.

Trace elements can also cause deleterious effects to salmonids, including olfactory stimulation, avoidance, and toxicity. Limited data (mostly from 1999–2002) indicate that most analyzed trace elements likely are not of concern; however, eight exceeded or were within an order of magnitude of water-quality criteria set to protect aquatic life: aluminum, cadmium, copper, iron, nickel, selenium, silver, and zinc. Factors confounding the trace element analysis include (1) data are not current, (2) sample counts were low (n = 1–2) at most sites, (3) samples were not collected throughout the year or at the same time of year, (4) water-quality criteria depend on water hardness, for which data are not available for many trace element samples, and (5) most trace element data represent the total recoverable rather than the dissolved fraction.

The lack of measurements of pesticide and trace element concentrations throughout the year with regular periodicity makes it difficult to assess potential impacts to salmonids, which are present year-round in the basin. Monitoring for particle-bound pesticides would also provide useful information. Analysis of pesticides that are commonly used in the basin but have not been analyzed for this project could be used to determine environmentally relevant mixtures of pesticides for future pesticide exposure studies on salmonids.

References Cited

Agency for Toxic Substances and Disease Registry, 1996, Toxicological profile for endrin: 191 p., accessed August 23, 2010, at http://www.atsdr.cdc.gov/ToxProfiles/tp89.pdf.

Amweg, E.L., Weston, D.P., and Ureda, N.M., 2005, Use and toxicity of pyrethroid pesticides in the Central Valley, California, USA: Environmental Toxicology and Chemistry, v. 24, no. 4, p. 966–972.

Amweg, E.L., Weston, D.P., You, J., and Lydy, M.J., 2006, Pyrethroid insecticides and sediment toxicity in urban creeks from California and Tennessee: Environmental Science and Technology, v. 40, p. 1700–1706.

BASF Corporation, 2010, Stamina® Material safety data sheet: Accessed June 10, 2010, from http://www.agproducts.basf.com/app/cdms?manuf=16&pd=9226&ms=2274.

Beketov, M.A., and Liess, M., 2008, Potential of 11 pesticides to initiate downstream drift of stream macroinvertebrates: Archives of Environmental Contamination and Toxicology, v. 55, p. 247–253.

Brander, S.M., Werner, Inge, White, J.W., and Deanovic, L.A., 2009, Toxicity of a dissolved pyrethroid mixture to *Hyalella azteca* at environmentally relevant concentrations: Environmental Toxicology and Chemistry, v. 28, no. 7, p. 1493–1499.

Buhl, K.J., and Hamilton, S.J., 1991, Relative sensitivity of early life stages of Arctic grayling, coho salmon, and rainbow trout to nine inorganics: Ecotoxicology and Environmental Safety, v. 22, p. 184–197.

California Environmental Protection Agency (CEPA), 1999, CEPA Office of Environmental Health Hazard Assessment Public health goal for endrin in drinking water: 21 p., accessed June 10, 2010, at http://oehha.ca.gov/water/phg/pdf/endrin_f.pdf.

California Environmental Protection Agency (CEPA), 2010, CEPA Department of Pesticide Regulation 2009 Status report pesticide contamination prevention act Annual Report: 25 p., accessed August 23, 2010, at http://www.cdpr.ca.gov/docs/emon/pubs/ehapreps/report_pcpa09.pdf.

Coats, J.R., and O'Donnell-Jeffery, N.L., 1979, Toxicity of four synthetic pyrethroid insecticides to rainbow trout: Bulletin of Environmental Contamination and Toxicology, v. 23, p. 250–255.

Coccoli, Holly, 2004, Hood River subbasin plan including lower Oregon Columbia Gorge tributaries: submitted to the Northwest Power and Conservation Planning Council by the Hood River Soil and Water Conservation District, 226 p.

Crossland, N.O., 1990, A review of the fate and toxicity of 3,4-dichloroaniline in aquatic environments: Chemosphere, v. 21, no. 12, p. 1489–1497.

Domagalski, J.L., Weston, D.P., Zhang, M., and Hladik, M., 2010, Pyrethroid insecticide concentrations and toxicity in streambed sediments and loads in surface waters of the San Joaquin Valley, California, USA: Environmental Toxicology and Chemistry, v. 29, no. 4, p. 813–823.

Ebbert, J.C., and Embrey, S.S., 2002, Pesticides in surface water of the Yakima River Basin, Washington, 1999–2000—Their occurrence and an assessment of factors affecting concentrations and loads: U.S. Geological Survey Water-Resources Investigations Report 01–4211, 49 p.

Gowan Company, 2004, Material safety data sheet for Imidan 50® WP: 6 p., accessed June 10, 2010, at http://www.gowanintl.com/Reference/Document.aspx?rid=7.

Grafton-Cardwell, E.E., Godfrey, L.D., Chaney, W.E., and Bentley, W.J., 2005, Various novel insecticides are less toxic to humans, more specific to key pests: California Agriculture, v. 59, no. 1, p. 29–34.

Hallberg, M., Renman, G., and Lundbom, T., 2007, Seasonal variations of ten metals in highway runoff and their partition between dissolved and particulate matter: Water, Air, and Soil Pollution, v. 181, p. 183–191.

Hecht, S.A., Baldwin, D.H., Mebane, C.A., Hawkes, T., Gross, S.J., and Sholz, N.L., 2007, An overview of sensory effects on juvenile salmonids exposed to dissolved copper—Applying a benchmark concentration approach to evaluate sublethal neurobehavioral toxicity: U.S. Department of Commerce NOAA Technical Memorandum NMFS-NWFSC-83, 39 p.

Henny, C.J., Grove, R.A., and Kaiser, J.L., 2008, Osprey distribution, abundance, reproductive success and contaminant burdens along lower Columbia River, 1997/1998 versus 2004: Archives of Environmental Contamination and Toxicology, v. 54, no. 3, p. 525–534.

Hinck, J.E., Schmitt, C.J., Bartish, T.M., Denslow, N.D., Blazer, V.S., Anderson, P.J., Coyle, J.J., Dethloff, G.M., and Tillitt, D.E., 2004, Biomonitoring of environmental status and trends (BEST) program—Environmental contaminants and their effects on fish in the Columbia River basin: U.S. Geological Survey Scientific Investigations Report 2004–5154, 125 p., http://www.cerc.usgs.gov/pubs/center/pdfDocs/BEST-Columbia_River.pdf.

Hollingsworth, C.S., ed., 2009, 2009 Pacific Northwest insect management handbook: Corvallis, Oregon State University Extension Service, 698 p.

Holmes, R.W., Anderson, B.S., Phillips, B.M., Hunt, J.W., Crane, D.B., Mekebri, Abdou, and Conner, Valerie, 2008, Statewide investigation of the role of pyrethroid pesticides in sediment toxicity in California's urban waterways: Environmental Science and Technology, v. 42, no. 18, p. 7003–7009.

Hood River Watershed Group, 2008, Hood River watershed action plan update: accessed June 4, 2010, at http://www.hoodriverswcd.org/HRWG/HRWatershedActionPlan.pdf.

Jaensson, Alia, Scott, A.P., Moore, Andrew, Kylin, Henrik, and Olsen, K.H., 2007, Effects of a pyrethroid pesticide on endocrine responses to female odours and reproductive behaviour in male parr of brown trout (*Salmo trutta* L.): Aquatic Toxicology, v. 81, no. 1, p. 1–9.

Jarrard, H.E., Delaney, K.R., and Kennedy, C.J., 2004, Impacts of carbamate pesticides on olfactory neurophysiology and cholinesterase activity in coho salmon (*Oncorhynchus kisutch*): Aquatic Toxicology, v. 69, p. 133–148.

Jenkins, J.J., 2003, Environmental monitoring of chlorpyrifos and azinphos-methyl dissolved residues in Hood River tributaries, Final Report to Hood River Soil and Water Conservation District: Corvallis, Oregon State University Agricultural Chemistry Research and Extension, 25 p.

Johnson, L.L., Ylitalo, G.M., Sloan, C.A., Anulacion, B.F., Kagley, A.N., Arkoosh, M.R., Lundrigan, T.A., Larson, K., Siipola, M., and Collier, T.K., 2007, Persistent organic pollutants in outmigrant juvenile Chinook salmon from the Lower Columbia Estuary, USA: Science of the Total Environment, v. 374, p. 342–366.

Julin, A.M., and Sanders, H.O., 1977, Toxicity and accumulation of the insecticide imidan in freshwater invertebrates and fishes: Transactions of the American Fisheries Society, v. 106, no. 4, p. 386–392.

Key, Peter, Chung, Katy, Siewicki, Tom, and Fulton, Mike, 2007, Toxicity of three pesticides individually and in mixture to larval grass shrimp (*Palaemonetes pugio*): Ecotoxicology and Environmental Safety, v. 68, no. 2, p. 272–277.

Kumaraguru, A.K., and Beamish, F.W.H., 1981, Lethal toxicity of permethrin (NRDC-143) to rainbow trout, *Salmo gairdneri*, in relation to body weight and water temperature: Water Research, v. 15, no. 4, p. 503–505.

Laetz, C.A., Baldwin, D.H., Collier, T.K., Hebert, V., Stark, J.D., and Scholz, N.L., 2009, The synergistic toxicity of pesticide mixtures—Implications for risk assessment and the conservation of endangered Pacific salmon: Environmental Health Perspectives, v. 117, no. 3, p. 348–353.

Lower Columbia River Estuary Partnership, 2007, Lower Columbia River and estuary ecosystem monitoring—Water quality and salmon sampling report: Portland, Oregon, 70 p., accessed August 30, 2010, at http://www.lcrep.org/sites/default/files/pdfs/WaterSalmonReport.pdf.

Lydy, M.J., and Austin, K.R., 2004, Toxicity assessment of pesticide mixtures typical of the Sacramento–San Joaquin Delta using *Chironomus tentans*: Archives of Environmental Contamination and Toxicology, v. 48, p. 49–55.

Marking, L.L., 1974, Toxicity of the synthetic pyrethroid SBP-1382 to fish: The Progressive Fish Culturist. v. 36, p. 144.

Mayer, F.L., and Ellersieck, M.R., 1988, Experiences with single-species tests for acute toxic effects on freshwater animals: Ambio, v. 17, no. 6, p. 367–375.

Michael, J.L., Webber, Jr., E.C., Bayne, D.R., Fischer, J.B., Gibbs, H.L., and Seesock, W.C., 1999, Hexazinone dissipation in forest ecosystems and impacts on aquatic communities: Canadian Journal of Forest Research, v. 29, p. 1170–1181.

Moore, Andrew, and Lower, Nicola, 2001, The impact of two pesticides on olfactory-mediated endocrine function in mature male Atlantic salmon (*Salmo salar L.*) parr: Comparative Biochemistry and Physiology Part B: Biochemistry and Molecular Biology, v. 129, no. 2–3, p. 269–276.

Munn, S.J., Aschberger, K., Cosgrove, O., Pakalin, S., Paya-Perez, A., Schwarz-Schulz, B., and Vegro, S., eds., 2006, European Union Risk Assessment Report: 3,4-dichloroaniline: Luxembourg, Office for Official Publications of the European Communities, 17 p.

National Marine Fisheries Service, 2008, Endangered Species Act Section 7 Consultation Biological Opinion Environmental Protection Agency registration of pesticides containing chlorpyrifos, diazinon, and malathion: 403 p., accessed April 28, 2010, at http://www nmfs.noaa.gov/pr/pdfs/pesticide_biop.pdf.

National Marine Fisheries Service, 2009, Endangered Species Act Section 7 Consultation Biological Opinion Environmental Protection Agency registration of pesticides containing carbaryl, carbofuran, and methomyl: 501 p., accessed April 28, 2010, at http://www nmfs.noaa.gov/pr/pdfs/carbamate.pdf.

National Pesticide Information Center (NPIC), 2003, NPIC Carbaryl technical fact sheet: 7 p., accessed June 9, 2010, at http://npic.orst.edu/factsheets/carbtech.pdf.

National Pesticide Information Center (NPIC), 2008, NPIC DEET technical fact sheet: 9 p., accessed June 9, 2010, at http://npic.orst.edu/factsheets/DEETtech.pdf.

National Pesticide Information Center (NPIC), 2010, NPIC Imidacloprid technical fact sheet: 14 p., accessed June 9, 2010, at http://npic.orst.edu/factsheets/imidacloprid.pdf.

New York State Department of Environmental Conservation (NYSDEC), 2004, NYSDEC Division of Solid and Hazardous Materials Letter to Judy Fersch, BASF Corporation: accessed June 2, 2010, at http://pmep.cce.cornell.edu/profiles/fung-nemat/febuconazole-sulfur/pyraclostrobin/pyraclos_let_1204 html.

Nieves-Puigdoller, Katherine, Björnsson, B.T., and McCormick, S.D., 2007, Effects of hexazinone and atrazine on the physiology and endocrinology of smolt development in Atlantic salmon: Aquatic Toxicology, v. 84, no. 1, p. 27–37.

Nilsen, E.B., Rosenbauer, R.R., Furlong, E.T., Burkhardt, M.R., Werner, S.L., Greaser, L., and Noriega, M., 2007, Pharmaceuticals, personal care products and anthropogenic waste indicators detected in streambed sediments of the Lower Columbia River and selected tributaries, *in* 6th International Conference on Pharmaceuticals and Endocrine Disrupting Chemicals in Water: Costa Mesa, CA, National Ground Water Association, Paper 4483, p. 15.

Oregon Department of Agriculture (ODA), 2008, ODA Pesticide Use Reporting System 2007 Annual Report, 40 p.

Oregon Department of Agriculture (ODA), 2009, ODA Pesticide Use Reporting System 2008 Annual Report, 32 p.

Oregon Department of Agriculture, Oregon Department of Environmental Quality, Oregon Department of Forestry, and Oregon Department of Human Services, 2008, Oregon State Pesticide Management Plan for Water Quality Protection, 42 p.

Oregon Department of Environmental Quality, 2004, Table 20 – Water quality criteria summary: accessed May 14, 2010, at http://www.deq.state.or.us/wq/rules/div041/table20.pdf.

Oregon Department of Environmental Quality (ODEQ), 2006, ODEQ Water Quality Division, Assessment methodology for Oregon's 2004/2006 integrated report on water quality status: 50 p., accessed July 26, 2010, at http://www.deq.state.or.us/wq/assessment/docs/methodology0406.pdf.

Oregon Department of Environmental Quality, 2008, Laboratory Analytical Storage and Retrieval (LASAR) database—LASAR web application, version 1.35: accessed August 9, 2010, at http://deq12.deq.state.or.us/LASAR2/.

Oregon State University, 1996a, Extension Toxicology Network Pesticide information profile for azinphos-methyl: accessed May 27, 2010, at http://extoxnet.orst.edu/pips/azinopho htm.

Oregon State University, 1996b, Extension Toxicology Network Pesticide information profile for chlorpyrifos: accessed May 27, 2010, at http://extoxnet.orst.edu/pips/chlorpyr htm.

Oregon State University, 1996c, Extension Toxicology Network Pesticide information profile for diuron: accessed May 27, 2010, at http://extoxnet.orst.edu/pips/diuron.htm.

Oregon State University, 1996d, Extension Toxicology Network Pesticide information profile for hexazinone: accessed May 27, 2010, at http://extoxnet.orst.edu/pips/hexazin htm.

Oregon State University, 1996e, Extension Toxicology Network Pesticide information profile for imidacloprid: accessed May 27, 2010, at http://extoxnet.orst.edu/pips/imidaclo.htm.

Oregon State University, 1996f, Extension Toxicology Network Pesticide information profile for malathion: accessed May 27, 2010, at http://extoxnet.orst.edu/pips/ malathio.htm.

Oregon State University, 1996g, Extension Toxicology Network Pesticide information profile for methomyl: accessed May 27, 2010, at http://extoxnet.orst.edu/pips/ methomyl htm.

Oregon State University, 1996h, Extension Toxicology Network Pesticide information profile for phosmet: accessed May 27, 2010, at http://extoxnet.orst.edu/pips/ phosmet.htm.

Oregon State University, 1996i, Extension Toxicology Network Pesticide information profile for simazine: accessed May 27, 2010, at http://extoxnet.orst.edu/pips/ simazine htm.

Oregon State University Extension Service, 2010, 2010 Pest management guide for tree fruits in the mid-Columbia area: accessed May 10, 2010, at http://ir.library.oregonstate.edu/ jspui/bitstream/1957/14105/1/EM8203.pdf.

Oropesa, A.L., García-Cambero, J.P., Gómez, L., Roncero, V., and Soler, F., 2009, Effect of long-term exposure to simazine on histopathology, hematological, and biochemical parameters in *Cyprinus carpio*: Environmental Toxicology, v. 24, no. 2, p. 187–199.

Peachey, Ed, ed., 2009, 2009 Pacific Northwest weed management handbook: Corvallis, Oregon, Oregon State University Extension Service, 543 p.

Pscheidt, J.W., and Ocamb, C.M., eds., 2009. 2009 Pacific Northwest plant disease management handbook: Corvallis, Oregon, Oregon State University Extension Service, 660 p.

Sandahl, J.F., Baldwin, D.H., Jenkins, J.J., and Scholz, N.L., 2005, Comparative thresholds for acetylcholinesterase inhibition and behavioral impairment in coho salmon exposed to chlorpyrifos: Environmental Toxicology and Chemistry, v. 24, no. 1, p. 136–145.

Scholz, N.L., Truelove, N.K., Labenia, J.S., Baldwin, D.H., and Collier, T.K., 2006, Dose-additive inhibition of Chinook salmon acetylcholinesterase activity by mixtures of organophosphate and carbamate insecticides: Environmental Toxicology and Chemistry, v. 25, no. 5, p. 1200–1207.

Schuler, L.J., Trimble, A.J., Belden, J.B., and Lydy, M.J., 2005, Joint toxicity of triazine herbicides and organophosphate insecticides to the midge *Chironomus tentans*: Archives of Environmental Contamination and Toxicology, v. 49, p. 173–177.

Sheedy, B.R., Lazorchak, J.M., Grunwald, D.J., Pickering, Q.H., Pilli, A., Hall, D., and Webb, R., 1991, Effects of pollution on freshwater organisms: Research Journal of the Water Pollution Control Federation, v. 63, no. 4, p. 619–696.

Sherman, T.J., Siipola, M.D., Abney, R.A., Ebner, D.B., Clarke, Joan, Ray, Gary, and Steevens, J.A., 2009, *Corbicula fluminea* as a bioaccumulation indicator species—A case study at the Columbia and Willamette rivers: Washington, D.C., U.S. Army Corps of Engineers ERDC/EL TR-09-3, 65 p., accessed September 2, 2010, at http://el.erdc.usace.army mil/elpubs/pdf/trel09-3.pdf.

Spehar, R.L., Lemke, A.E., Pickering, Q.H., Roush, T.H., Russo, R.C., and Yount, J.D., 1981, Effects of pollution on freshwater fish: Journal of the Water Pollution Control Federation, v. 53, no. 6, p. 1028–1076.

StreamNet, 2010, Generalized fish distribution—All species combined: StreamNet ArcGIS geodatabase, accessed August 17, 2010, from http://www.streamnet.org/mapping_ apps.cfm.

Tierney, K.B., Baldwin, D.H., Hara, T.J., Ross, P.S., Scholz, N.I., and Kennedy, C.J., 2010, Olfactory toxicity in fishes: Aquatic Toxicology, v. 96, p. 2–26.

Trimble, A.J., and Lydy, M.J., 2006, Effects of triazine herbicides on organophosphate insecticide toxicity in *Hyalella azteca*: Archives of Environmental Contamination and Toxicology, v. 51, p. 29–34.

U.S. Department of Agriculture (USDA), 2007, USDA Census of Agriculture vol.1, ch.2, County level data, Table 42 – Fertilizers and chemicals applied 2007 and 2002, accessed March 17, 2011, at http://www.agcensus.usda. gov/Publications/2007/Full_Report/Volume_1,_Chapter_2_ County_Level/Oregon/st41_2_042_042.pdf.

U.S. Department of Agriculture (USDA), 2009, USDA Agricultural Research Service Pesticide properties database, accessed March 17, 2011, at http://www.ars.usda.gov/ Services/docs.htm?docid=14199.

U.S. Department of Agriculture Forest Service, 1996, Mt. Hood National Forest, West Fork of Hood River watershed analysis, Mt. Hood-Parkdale, OR: 284 p., accessed March 16, 2011, at http://www.fs.usda.gov/Internet/FSE_ DOCUMENTS/fsbdev3_036588.pdf.

U.S. Census Bureau, 2010, Table 4 – Annual estimates of the resident population for incorporated places in Oregon April 1, 2000 to July 1, 2009: accessed August 31, 2010, from http://www.census.gov/popest/cities/SUB-EST2009-4 html.

U.S. Environmental Protection Agency, 1994, U.S. Environmental Protection Agency Office of Prevention, Pesticides and Toxic Substances Reregistration eligibility decision (RED) for hexazinone: 60 p., accessed May 14, 2010, at http://www.epa.gov/oppsrrd1/REDs/0266.pdf.

U.S. Environmental Protection Agency, 1996, U.S. Environmental Protection Agency Office of Prevention, Pesticides and Toxic Substances Reregistration eligibility decision (RED) for norflurazon: 98 p., accessed May 14, 2010, at http://www.epa.gov/oppsrrd1/REDs/0229.pdf.

U.S. Environmental Protection Agency, 1997, U.S. Environmental Protection Agency Office of Prevention, Pesticides and Toxic Substances Reregistration eligibility decision (RED) for propoxur: 64 p., accessed May 14, 2010, at http://www.epa.gov/oppsrrd1/REDs/2555red.pdf.

U.S. Environmental Protection Agency, 1998a, U.S. Environmental Protection Agency Office of Prevention, Pesticides and Toxic Substances Reregistration eligibility decision (RED) for DEET: 42 p., accessed May 14, 2010, at http://www.epa.gov/oppsrrd1/REDs/0002red.pdf.

U.S. Environmental Protection Agency, 1998b, U.S. Environmental Protection Agency Office of Prevention, Pesticides and Toxic Substances Reregistration eligibility decision (RED) for methomyl: 138 p., accessed May 14, 2010, at http://www.epa.gov/oppsrrd1/REDs/0028red.pdf.

U.S. Environmental Protection Agency, 1999, U.S. Environmental Protection Agency Office of Prevention, Pesticides and Toxic Substances Reregistration eligibility decision (RED) for chlorothalonil: 216 p., accessed June 8, 2010, at http://www.epa.gov/oppsrrd1/REDs/0097red.pdf.

U.S. Environmental Protection Agency, 2000, Status of chemicals in special review: EPA-738-R-00-001, 59 p., accessed July 31, 2010, at http://www.epa.gov/oppsrrd1/special_review/sr00status.pdf.

U.S. Environmental Protection Agency, 2001, U.S. Environmental Protection Agency Office of Pesticide Programs Phosmet Fact Sheet: accessed May 14, 2010, at http://www.epa.gov/oppsrrd1/REDs/factsheets/phosmet_fs.htm.

U.S. Environmental Protection Agency, 2003, U.S. Environmental Protection Agency Office of Prevention, Pesticides and Toxic Substances Reregistration eligibility decision (RED) for diuron: 106 p., accessed May 14, 2010, at http://www.epa.gov/oppsrrd1/reregistration/REDs/diuron_red.pdf.

U.S. Environmental Protection Agency, 2004, Availability of court orders in Washington Toxics Coalition v. EPA litigation: Federal Register, v. 69, no. 31, p. 7478–7480.

U.S. Environmental Protection Agency, 2005a, National recommended water-quality criteria table, accessed May 14, 2010, at http://water.epa.gov/scitech/swguidance/waterquality/standards/current/index.cfm.

U.S. Environmental Protection Agency, 2005b, U.S. Environmental Protection Agency Office of Pesticide Programs Environmental Effects Division Pesticide ecological effects database guidance manual: 11 p., accessed August 30, 2010 from http://www.ipmcenters.org/Ecotox/index.cfm.

U.S. Environmental Protection Agency, 2005c, U.S. Environmental Protection Agency Office of Pesticide Programs Revised environmental fate and ecological risk assessment of fluometuron: 54 p., accessed May 14, 2010, from http://www.regulations.gov/search/Regs/home.html#docketDetail?R=EPA-HQ-OPP-2004-0372.

U.S. Environmental Protection Agency, 2005d, U.S. Environmental Protection Agency Office of Prevention, Pesticides and Toxic Substances Overview of the use and usage of soil fumigants: 30 p., accessed January 10, 2011, at http://www.epa.gov/oppsrrd1/reregistration/x-soil-fum-HOLD/soil_fumigant_use.pdf.

U.S. Environmental Protection Agency, 2005e, U.S. Environmental Protection Agency Office of Prevention, Pesticides and Toxic Substances Reregistration eligibility decision (RED) for fluometuron: 44 p., accessed May 14, 2010, at http://www.epa.gov/oppsrrd1/REDs/fluometuron_red.pdf.

U.S. Environmental Protection Agency, 2006a, U.S. Environmental Protection Agency Office of Prevention, Pesticides and Toxic Substances Reregistration eligibility decision (RED) for azinphos-methyl: 130 p., accessed May 14, 2010, at http://www.epa.gov/oppsrrd1/reregistration/REDs/azm_red.pdf.

U.S. Environmental Protection Agency, 2006b, U.S. Environmental Protection Agency Office of Prevention, Pesticides and Toxic Substances Decision documents for atrazine: 126 p., accessed May 14, 2010, at http://www.epa.gov/oppsrrd1/REDs/atrazine_combined_docs.pdf.

U.S. Environmental Protection Agency, 2006c, U.S. Environmental Protection Agency Office of Prevention, Pesticides and Toxic Substances Reregistration eligibility decision (RED) for chlorpyrifos: 114 p., accessed May 14, 2010, at http://www.epa.gov/oppsrrd1/reregistration/REDs/chlorpyrifos_red.pdf.

U.S. Environmental Protection Agency, 2006d, U.S. Environmental Protection Agency Office of Prevention, Pesticides and Toxic Substances Reregistration eligibility decision (RED) for diazinon: 68 p., accessed May 14, 2010, at http://www.epa.gov/oppsrrd1/reregistration/REDs/diazinon_red.pdf.

U.S. Environmental Protection Agency, 2006e, U.S. Environmental Protection Agency Office of Prevention, Pesticides and Toxic Substances Reregistration eligibility decision (RED) for phosmet: 110 p., accessed May 14, 2010, at http://www.epa.gov/oppsrrd1/reregistration/REDs/phosmet_red.pdf.

U.S. Environmental Protection Agency, 2006f, U.S. Environmental Protection Agency Office of Prevention, Pesticides and Toxic Substances Reregistration eligibility decision (RED) for propiconazole: 86 p., accessed May 14, 2010, at http://www.epa.gov/oppsrrd1/REDs/propiconazole_red.pdf.

U.S. Environmental Protection Agency, 2006g, U.S. Environmental Protection Agency Office of Prevention, Pesticides and Toxic Substances Reregistration eligibility decision (RED) for simazine: 77 p., accessed May 14, 2010, at http://www.epa.gov/oppsrrd1/REDs/simazine_red.pdf.

U.S. Environmental Protection Agency, 2007, ECOTOX User Guide, ECOTOXicology Database System, version 4.0: accessed July 20, 2010, at http://www.epa.gov/ecotox/.

U.S. Environmental Protection Agency, 2008, U.S. Environmental Protection Agency Office of Pesticide Programs Diazinon fact sheet: 1 p., accessed June 1, 2010, at http://www.epa.gov/opp00001/factsheets/chemicals/diazinon-factsheet.htm.

U.S. Environmental Protection Agency, 2009a, U.S. Environmental Protection Agency Office of Groundwater and Drinking Water Technical factsheet on endrin: 3 p., accessed June 1, 2010, at http://www.epa.gov/safewater/pdfs/factsheets/soc/tech/endrin.pdf.

U.S. Environmental Protection Agency, 2009b, U.S. Environmental Protection Agency Office of Pesticide Programs' Aquatic life benchmarks: accessed May 14, 2010, at http://www.epa.gov/oppefed1/ecorisk_ders/aquatic_life_benchmark.htm.

U.S. Environmental Protection Agency, 2009c, U.S. Environmental Protection Agency Office of Pesticide Programs Azinphos-methyl phaseout: accessed July 16, 2010, at http://www.epa.gov/oppsrrd1/reregistration/azm/phaseout_fs.htm.

U.S. Environmental Protection Agency, 2009d, U.S. Environmental Protection Agency Office of Prevention, Pesticides and Toxic Substances Revised reregistration eligibility decision (RED) for malathion: 139 p., accessed May 14, 2010, at http://www.epa.gov/oppsrrd1/REDs/malathion-red-revised.pdf.

U.S. Environmental Protection Agency, 2010a, U.S. Environmental Protection Agency Office of Chemical Safety and Pollution Prevention Letter to James H. Lecky, Director, Office of Protected Resources, National Marine Fisheries Service, May 14, 2010: accessed June 1, 2010, at http://www.epa.gov/espp/litstatus/wtc/biop-ltr-to-jhlecky-may-2010.pdf.

U.S. Environmental Protection Agency, 2010b, U.S. Environmental Protection Agency Office of Pesticide Programs Technical overview of ecological risk assessment – analysis phase, ecological effects characterization, accessed September 2, 2010, at http://www.epa.gov/oppefed1/ecorisk_ders/toera_analysis_eco.htm.

U.S. Fish and Wildlife Service, 2010, Endangered Species Program home page: accessed June 3, 2010, at http://www.fws.gov/endangered/.

U.S. Geological Survey, 2004, USGS Columbia Environmental Research Center Acute toxicity database, accessed August 3, 2010, at http://www.cerc.usgs.gov/data/acute/acute.html.

U.S. Geological Survey, 2010, National water information system database, accessed July 31, 2010, at: http://nwis.waterdata.usgs.gov/nwis.

Ural, M.Ş., and Sağlam, N., 2005, A study on the acute toxicity of pyrethroid deltamethrin on the fry rainbow trout (*Oncorhynchus mykiss* Walbaum, 1792): Pesticide Biochemistry and Physiology, v. 83, p. 124–131.

Velisek, J., Stastna, K., Sudova, E., Turek, J., and Svobodova, Z., 2009, Effects of subchronic simazine exposure on some biometric, biochemical, hematological, and histopathological parameters of common carp (*Cyprinus carpio L.*): Neuroendocrinology Letters, v. 30, Suppl. 1, p. 236–241.

Wan, M.T., Watts, R.G., and Moul, D.J., 1988, Evaluation of the acute toxicity to juvenile Pacific salmonids of hexazinone and its formulated products – Pronone 10G, Velpar® L, and their carriers: Bulletin of Environmental Contamination and Toxicology, v. 41, p. 609–616.

Weston, D.P., You, J., and Lydy, M.J., 2004, Distribution and toxicity of sediment-associated pesticides in agriculture-dominated water bodies of California's Central Valley: Environmental Science and Technology, v. 38, p. 2752–2759.

Appendixes A–J

Appendix A. Catchment Area and Upstream Land Use for Water Sampling Sites in the Hood River Basin, Oregon.

[**Abbreviations:** ODEQ, Oregon Department of Environmental Quality; ODFW, Oregon Department of Fish and Wildlife; km^2, square kilometers; RM, river mile; Ppl, Pacific Power and Light]

Map ID	ODEQ station ID	Site name (full)	Site name (short)	Catchment area (km^2)	Percentage[1]					
					Forest	Urban	Orchard	Non-orchard agriculture	Water	Other
1	13181	Baldwin Creek at end of Baldwin Creek Road	Baldwin	15.01	73	1	17	8	0	1
2	25133	Dog River below Puppy Creek confluence	Dog	32.76	99	0	0	0	0	0
3	25124	Evans Creek at bridge (Baseline Road)	Evans	5.81	70	0	15	14	1	1
4	13138	East Fork Hood River at County Gravel Pit (River Mile 0.75)	Hood, East Fork	265.9	80	0	6	6	2	6
5	13139	Middle Fork Hood River at River Mile 1.0 (ODFW Smolt Trap)	Hood, Middle Fork	106.39	89	0	1	3	2	6
6	13158	Hood River downstream of Ppl Powerdale Powerhouse	Hood, mouth	879.41	81	0	8	6	1	3
7	12012	Hood River at footbridge downstream of I-84	Hood, mouth	879.41	81	0	8	6	1	3
8	10681	West Fork Hood River at mouth	Hood, West Fork, mouth	264.93	96	0	0	1	1	1
9	34787	West Fork Hood River at Moving Falls (RM 2.5)	Hood, West Fork, RM 2.5	264.93	96	0	0	1	1	2
10	13140	West Fork Hood River at Lost Lake Road (River Mile 4.7)	Hood, West Fork, RM 4.7	178.55	96	0	0	1	2	2
11	21634	Indian Creek near mouth	Indian	16.86	19	4	40	34	0	1
12	11972	Lenz Creek at mouth	Lenz	8.63	12	1	56	26	0	1
13	31499	Middle Neal Creek at Hwy 35	Neal, middle	66.77	89	0	6	4	0	1
14	13141	Neal Creek at mouth (upstream of bridge)	Neal, mouth	85.93	75	0	15	8	0	1
15	25123	Upper Neal Creek above agri-culture diversion	Neal, upper, above diversion	20.95	97	0	0	2	0	0
16	30174	Upper Neal Creek, downstream	Neal, upper, below diversion	52.88	95	0	1	2	0	1
17	34788	Rogers Spring Creek at Red Hill Driver (RM 0.25)	Rogers	0.58	9	0	4	10	0	76

[1]Percentages may not total 100 due to rounding or the exclusion of minor land use categories from this table.

Appendix B. Number of Samples Collected in the Hood River Basin, Oregon, 1999–2009, by Site, Month, and Year

[**Abbreviations:** –, no samples collected]

Site and year	Jan	Feb	Mar	Apr	May	Jun	Jul	Aug	Sept	Oct	Nov	Dec	Annual total
Baldwin	–	–	7	5	5	15	4	3	11	2	–	–	52
2003	–	–	–	4	–	3	4	3	3	2	–	–	19
2004	–	–	–	–	3	5	–	–	4	–	–	–	12
2005	–	–	7	1	1	5	–	–	4	–	–	–	18
2006	–	–	–	–	1	2	–	–	–	–	–	–	3
Dog	–	–	9	11	3	10	4	–	4	–	–	–	41
2001	–	–	8	1	–	2	–	–	–	–	–	–	11
2002	–	–	1	6	–	–	–	–	–	–	–	–	7
2003	–	–	–	4	–	3	4	–	–	–	–	–	11
2004	–	–	–	–	3	5	–	–	4	–	–	–	12
Evans	–	–	18	13	5	21	10	3	11	2	–	–	83
2001	–	–	9	2	–	3	–	–	–	–	–	–	14
2002	–	–	1	6	–	2	6	–	–	–	–	–	15
2003	–	–	1	4	–	3	4	3	3	2	–	–	20
2004	–	–	–	–	3	5	–	–	4	–	–	–	12
2005	–	–	7	1	1	6	–	–	4	–	–	–	19
2006	–	–	–	–	1	2	–	–	–	–	–	–	3
Hood, East Fork	–	1	9	10	1	4	6	–	–	1	–	–	32
1999	–	–	–	–	–	–	–	–	–	1	–	–	1
2000	–	1	8	4	1	2	–	–	–	–	–	–	16
2002	–	–	1	6	–	2	6	–	–	–	–	–	15
Hood, Middle Fork	–	–	8	4	1	3	–	–	–	1	–	–	17
1999	–	–	–	–	–	–	–	–	–	1	–	–	1
2000	–	–	8	4	1	3	–	–	–	–	–	–	16
Hood, mouth	–	1	24	18	15	18	–	–	9	3	1	1	90
1999	–	–	3	–	–	–	–	–	–	1	–	–	4
2000	–	1	1	–	2	1	–	–	–	–	–	–	5
2001	–	–	4	2	1	4	–	–	–	–	–	–	11
2002	–	–	4	3	–	–	–	–	–	–	–	–	7
2005	–	–	5	–	–	6	–	–	4	–	–	–	15
2006	–	–	–	–	1	2	–	–	4	–	–	–	7
2007	–	–	2	3	4	1	–	–	–	1	1	1	13
2008	–	–	3	5	4	–	–	–	–	–	–	–	12
2009	–	–	2	5	3	4	–	–	1	1	–	–	16
Hood, West Fork, mouth	–	–	4	9	3	3	–	–	1	1	–	–	21
2008	–	–	2	4	–	–	–	–	–	–	–	–	6
2009	–	–	2	5	3	3	–	–	1	1	–	–	15
Hood, West Fork, RM 2.5	–	–	–	1	4	1	–	–	–	–	–	–	6
2008	–	–	–	1	4	–	–	–	–	–	–	–	5
2009	–	–	–	–	–	1	–	–	–	–	–	–	1
Hood, West Fork, RM 4.7	–	–	12	4	1	2	–	–	–	1	–	–	20
1999	–	–	3	–	–	–	–	–	–	1	–	–	4
2000	–	–	8	4	1	2	–	–	–	–	–	–	15
2001	–	–	1	–	–	–	–	–	–	–	–	–	1
Indian	–	–	4	–	1	–	–	–	–	–	–	–	5
1999	–	–	3	–	–	–	–	–	–	–	–	–	3
2000	–	–	1	–	1	–	–	–	–	–	–	–	2

Appendix B.   Number of samples collected in the Hood River basin, Oregon, 1999–2009, by site, month, and year.—Continued

[**Abbreviations:** –, no samples collected]

Site and year	Jan	Feb	Mar	Apr	May	Jun	Jul	Aug	Sept	Oct	Nov	Dec	Annual total
Lenz	–	–	29	13	15	35	2	3	13	4	–	–	114
2001	–	–	9	1	2	5	–	–	–	–	–	–	17
2002	–	–	6	2	1	5	2	–	–	–	–	–	16
2003	–	–	2	2	–	8	–	3	3	2	–	–	20
2004	–	–	5	–	3	5	–	–	4	–	–	–	17
2005	–	–	5	–	1	6	–	–	5	1	–	–	18
2006	–	–	–	–	1	2	–	–	–	–	–	–	3
2008	–	–	–	3	4	–	–	–	–	–	–	–	7
2009	–	–	2	5	3	4	–	–	1	1	–	–	16
Neal, middle	–	–	12	13	13	18	–	–	13	2	1	1	73
2004	–	–	–	–	–	5	–	–	4	–	–	–	9
2005	–	–	5	–	1	6	–	–	4	–	–	–	16
2006	–	–	–	–	1	2	–	–	4	–	–	–	7
2007	–	–	2	3	4	1	–	–	–	1	1	1	13
2008	–	–	3	5	4	–	–	–	–	–	–	–	12
2009	–	–	2	5	3	4	–	–	1	1	–	–	16
Neal, mouth	–	1	45	23	22	41	3	3	17	5	1	1	162
1999	–	–	3	–	–	–	–	–	–	1	–	–	4
2000	–	1	8	4	2	6	1	–	–	–	–	–	22
2001	–	–	8	1	2	5	–	–	–	–	–	–	16
2002	–	–	6	3	1	5	2	–	–	–	–	–	17
2003	–	–	3	2	1	7	–	3	3	2	–	–	21
2004	–	–	5	–	3	5	–	–	4	–	–	–	17
2005	–	–	5	–	1	6	–	–	5	–	–	–	17
2006	–	–	–	–	1	2	–	–	4	–	–	–	7
2007	–	–	2	3	4	1	–	–	–	1	1	1	13
2008	–	–	3	5	4	–	–	–	–	–	–	–	12
2009	–	–	2	5	3	4	–	–	1	1	–	–	16
Neal, upper, above diversion	–	–	32	9	13	31	6	3	15	3	1	–	113
2001	–	–	9	1	2	5	–	–	–	–	–	–	17
2002	–	–	6	2	1	5	6	–	–	–	–	–	20
2003	–	–	3	2	1	7	–	3	3	2	–	–	21
2004	–	–	5	–	3	5	–	–	4	–	–	–	17
2005	–	–	7	1	1	6	–	–	4	–	–	–	19
2006	–	–	–	–	1	2	–	–	4	–	–	–	7
2007	–	–	2	3	4	1	–	–	–	1	1	–	12
Neal, upper, below diversion	–	–	15	15	17	25	–	3	16	4	1	1	97
2003	–	–	3	2	1	7	–	3	3	2	–	–	21
2004	–	–	–	–	3	5	–	–	4	–	–	–	12
2005	–	–	5	–	1	6	–	–	4	–	–	–	16
2006	–	–	–	–	1	2	–	–	4	–	–	–	7
2007	–	–	2	3	4	1	–	–	–	1	1	1	13
2008	–	–	3	5	4	–	–	–	–	–	–	–	12
2009	–	–	2	5	3	4	–	–	1	1	–	–	16
Rogers	–	–	5	10	7	3	–	–	1	1	–	–	27
2008	–	–	3	5	4	–	–	–	–	–	–	–	12
2009	–	–	2	5	3	3	–	–	1	1	–	–	15
Total - all sites and years	–	3	233	158	126	230	35	18	111	30	5	4	953

Appendix C. Pesticide Products Suitable for the Major Land Uses of the Hood River Basin, Oregon

[**Source:** Hollingsworth, 2009; Peachey, 2009; Pscheidt and Ocamb, 2009; Oregon State University Extension Service, 2010. **Abbreviations:** X, pesticide suitable for the listed application; – pesticide not suitable for the listed application, not analyzed in this project, or example product names not provided]

Pesticide	Product names	Blueberries	Forestry	Grapes	Household	Orchards	Pasture, hay, range	Rights-of-way	Years analyzed
Coddling moth mating disruption									
(Z)-I I-Tetradecen-I-yl Acetate	Nomate	–	–	–	–	X	–	–	–
E-11-Tetradecen-1-yl Acetate + (E,E)-9,11-Tetradecadien-1-yl Acetate	Isomate	–	–	–	–	X	–	–	–
MCPA ester	Checkmate	–	–	–	–	X	–	–	–
Products for disease control									
1,3 dichloropropene	Telone II	X	–	X	–	–	–	–	–
Azoxystrobin	Abound	X	–	X	–	–	–	–	–
Bicarbonate products	Armicarb, Kaligreen, MilStop, Monterey Bi-Carb	–	–	X	–	X	–	–	–
Calcium polysulfide	lime sulfur	X	–	–	X	X	–	–	–
Chloropicrin	–	–	–	–	–	X	–	–	–
Dazomet	Basamid G	X	–	X	–	–	–	–	–
Dichloran	Botran	–	–	X	–	–	–	–	–
Dimethyl phenol	Gallex	X	–	–	–	–	–	–	–
Iprodione	Iprodione	X	–	X	–	–	–	–	–
Kaolin	Surround	–	–	X	X	X	–	–	–
Mancozeb	–	–	–	X	–	X	–	–	–
Metam sodium	Vapam, Sectagon 42, Metam CLR	–	–	X	–	X	–	–	–
Methyl bromide	–	–	–	X	–	X	–	–	–
Methyl phenol	Gallex	X	–	–	–	–	–	–	–
Mono- and dipotassium salts of phosphorous acid	Agri-Fos	X	–	–	–	–	–	–	–
Monopotassium phosphite + dipotassium phosphite	Fosphite	X	–	–	–	X	–	–	–
Sulfur products[1]	–	–	–	X	X	X	–	–	–
Fungicides									
Boscalid	Endura, Pristine	X	–	X	–	X	–	–	–
Captan	Captan, Captec	X	–	X	–	X	–	–	–
Chlorothalonil	Bravo Weather Stik	–	–	–	X	X	–	–	2009
Copper products	–	X	–	X	X	X	–	–	2000–01
Cyprodinil	Vangard, Switch	X	–	X	–	–	–	–	–
Dodine	Syllit	–	–	–	–	X	–	–	–
Fenarimol	Rubigan	–	–	X	–	X	–	–	2009
Fenbuconazole	Indar	X	–	–	–	X	–	–	–
Fenhexamid	Elevate, CaptEvate	X	–	X	–	X	–	–	–
Fludioxinil	Switch	X	–	–	–	–	–	–	–
Fosetyl-aluminum	Aliette	X	–	–	–	X	–	–	–
Kresoxim-methyl	Sovarn	–	–	X	–	–	–	–	–
Metalaxyl	Ridomil Gold	X	–	–	–	X	–	–	–
Metconazole	Quash	–	–	–	–	X	–	–	–
Myclobutanil	–	–	–	X	–	X	–	–	–
Potassium bicarbonate	Remedy	–	–	–	X	–	–	–	–
Propiconazole	Tilt	–	–	–	X	X	–	–	2009

Appendix C.   Pesticide products suitable for the major land uses of the Hood River basin, Oregon.—Continued

[**Source:** Hollingsworth, 2009; Peachey, 2009; Pscheidt and Ocamb, 2009; Oregon State University Extension Service, 2010. **Abbreviations:** X, pesticide suitable for the listed application; – pesticide not suitable for the listed application, not analyzed in this project, or example product names not provided]

Pesticide	Product names	Blueberries	Forestry	Grapes	Household	Orchards	Pasture, hay, range	Rights-of-way	Years analyzed
Fungicides (Continued)									
Pyraclostrobin	–	X	–	–	–	X	–	–	2009
Pyrimethanil	Scala	–	–	X	–	–	–	–	–
Quinoxyfen	Quintec	–	–	X	–	X	–	–	–
Sodium, potassium, and ammonium phosphites	Phostrol	X	–	–	–	–	–	–	–
Sodium tatrathiocarbonate[1]	Enzone	–	–	X	–	–	–	–	–
Streptomycin	Agrimycin	–	–	–	–	X	–	–	–
Tebuconazole	Elite, Orius	–	–	X	X	X	–	–	–
Terramycin	Mycoshield	–	–	–	–	X	–	–	–
Thiophanate-methyl	–	–	–	X	X	X	–	–	–
Trifloxystrobin	–	–	–	X	–	X	–	–	–
Triflumizole	–	–	–	X	–	X	–	–	–
Triforine	Funginex	–	–	–	X	–	–	–	–
Ziram	–	X	–	X	–	X	–	–	–
Products to prevent fruit drop									
Aminoethoxyvinylglycine hydrochloride	Retain	–	–	–	–	X	–	–	–
Napthalene acetic acid (NAA)	NAA	–	–	–	–	X	–	–	–
Herbicides									
2,4-D	Crossbow, Curtail, Weedmaster, Pasturemaker, Cimarron Max	X	X	–	X	–	X	X	2009
2,4-D amine	Saber, Weed-Rhap A4d, Dri-Clean Herbicide	–	–	X	–	X	–	–	2009
2,4-D ester	Crossbow	–	–	–	–	–	–	X	2009
Aminopyralid	Milestone	–	–	–	X	–	X	–	–
Atrazine	–	–	X	–	–	–	–	–	1999–09
Bromacil	Krovar	–	–	–	–	–	–	X	2009
Carfentrazone	Aim	X	–	–	–	–	X	–	–
Chlorsulfuron	Telar	–	X	–	–	–	X	–	–
Clethodim	Envoy, Prism, Select	X	–	X	–	X	–	–	–
Clopyralid	–	–	X	–	X	X	X	–	–
Dicamba	Banvel, Vanquish, Clarity, Weedmaster, Pasturemaker, Latigo	–	–	–	X	–	X	–	–
Dichlobenil	Casoron	X	–	X	X	X	–	–	–
Diquat	Reglone	–	–	X	–	–	–	–	–
Diuron	–	X	–	X	–	X	–	X	2009
Fluazifop	Flusilade	X	–	X	–	X	–	–	–
Flumioxazin	Chateau	X	–	X	–	X	–	–	–
Fluroxypyr	Starane, Surmount, PastureGard	–	–	–	X	–	X	–	–
Glufosinate ammonium	Rely	X	–	X	–	X	–	–	–
Glyphosate	–	X	X	X	X	X	X	X	–
Halosulfuron	Sandea	–	–	–	–	X	–	–	–

Appendix C.   Pesticide products suitable for the major land uses of the Hood River basin, Oregon.—Continued

[**Source:** Hollingsworth, 2009; Peachey, 2009; Pscheidt and Ocamb, 2009; Oregon State University Extension Service, 2010. **Abbreviations:** X, pesticide suitable for the listed application; – pesticide not suitable for the listed application, not analyzed in this project, or example product names not provided]

Pesticide	Product names	Blueberries	Forestry	Grapes	Household	Orchards	Pasture, hay, range	Rights-of-way	Years analyzed
Herbicides (Continued)									
Hexazinone	–	X	X	–	–	–	X	–	2009
Imazapic	Plateau	–	X	–	–	–	X	–	–
Imazapyr	–	–	X	–	–	–	–	–	2009
Isoxaben	Gallery or Gallery T&V, Showcase, Snapshot	X	–	X	–	X	–	–	–
MCPA	–	–	–	–	X	–	X	–	–
Mesotrione	Callisto	X	–	–	–	–	–	–	–
Metsulfuron methyl	Cimarron Max, Escort	–	X	–	X	–	X	X	–
Napropamide	Devrinol	X	–	X	–	X	–	–	2009
Norflurazon	Solicam	X	–	X	–	X	–	–	2009
Oryzalin	–	X	–	X	–	X	–	–	–
Oxyfluorfen	–	X	–	X	–	X	–	–	–
Paraquat	Gramoxone Inteon, Firestorm, Cyclone	–	–	X	–	X	X	–	–
Pendimethalin	Prowl H2	–	–	X	–	X	–	–	2009
Picloram	–	–	X	–	–	–	X	–	–
Pronamide	–	X	–	X	–	X	–	–	2009
Rimsulfuron	Matrix	–	–	X	–	X	–	–	–
Sethoxydim	Poast	X	X	X	–	X	–	–	–
Simazine	–	X	–	X	–	X	–	–	2009
Sulfometuron methyl	–	–	X	–	–	–	–	X	–
Tebuthiuron	Spike	–	–	–	–	–	X	–	2009
Terbacil	Sinbar	X	–	–	–	X	–	–	2009
Triasulfuron	Amber	–	–	–	–	–	X	–	–
Triclopyr	–	–	X	–	X	–	X	X	2009
Triclopyr ester	–	–	X	–	–	–	X	–	–
Trifluralin	Showcase, Snapshot, Treflan	X	–	X	–	X	–	–	2009
Insecticides									–
Abamectin	–	–	–	X	–	X	–	–	–
Acephate	–	–	–	–	X	–	–	–	–
Acetamiprid	Assail	X	–	X	X	X	–	–	–
Azadirachtin	Azatin, Neemix	X	–	X	–	–	–	–	–
Azinphos methyl	Guthion	–	–	–	–	X	–	–	1999–09
Bifenazate	Acramite	–	–	X	–	–	–	–	–
Bifenthrin	Brigade	–	–	X	X	–	–	–	–
Buprofezin	Applaud, Centaur	–	–	X	–	X	–	–	–
Carbaryl	Sevin	–	–	X	X	X	X	–	1999–00, 2009
Chlorantraniliprole	Voliam Flexi	–	–	X	–	–	–	–	–
Chlorpyrifos	Lorsban	–	–	X	–	X	–	–	1999–09
Clothianidin	–	–	–	–	–	X	–	–	–
Cyfluthrin	Baythroid	–	–	–	X	–	X	–	–
Deltamethrin	–	–	–	–	–	X	–	–	–
Diazinon	Diazinon	X	–	–	–	X	–	–	1999–09
Dicofol	Kelthane	–	–	X	–	X	–	–	–
Diflubenzuron	Dimilin	–	–	–	–	–	X	–	–
Dimethoate	Dimethoate	–	–	–	–	X	–	–	1999–09
Disulfoton	–	–	–	–	X	–	–	–	2009

Appendix C.   Pesticide products suitable for the major land uses of the Hood River basin, Oregon.—Continued

[**Source:** Hollingsworth, 2009; Peachey, 2009; Pscheidt and Ocamb, 2009; Oregon State University Extension Service, 2010. **Abbreviations:** X, pesticide suitable for the listed application; – pesticide not suitable for the listed application, not analyzed in this project, or example product names not provided]

Pesticide	Product names	Blueberries	Forestry	Grapes	Household	Orchards	Pasture, hay, range	Rights-of-way	Years analyzed
Insecticides (Continued)									–
Dormant oil	–	–	–	–	X	–	–	–	–
Emamectin benzoate	Proclaim	–	–	–	–	X	–	–	–
Endosulfan	–	–	–	X	–	X	–	–	2009
Esfenvalerate	Asana	X	–	–	X	X	–	–	2003–09
Fenbutatin oxide	Vendex	–	–	X	X	X	–	–	–
Fenpropathrin	–	X	–	X	–	X	–	–	–
Gamma-cyhalothrin	–	–	–	–	–	X	–	–	–
Imidacloprid	–	X	–	X	X	X	–	–	2009
Indoxacarb	Avaunt	–	–	–	–	X	–	–	–
Insecticidal soap	M-Pede, others	X	–	X	X	–	–	–	–
Iron phosphate	–	–	–	–	X	–	–	–	–
Lambda-cyhalothrin	–	–	–	–	X	X	X	–	–
Malathion	Malathion	X	–	X	X	X	X	–	1999–09
Metaldehyde	–	–	–	–	X	–	–	–	–
Methidathion	Supracide	–	–	–	–	X	X	–	–
Methomyl	Lannate	X	–	X	–	–	–	–	2009
Methoxyfenozide	Intrepid	X	–	–	–	X	X	–	–
Methyl parathion	Methyl 4EC	–	–	–	–	–	X	–	1999–02, 2009
Novaluron	Rimon	–	–	–	–	X	–	–	–
Oxamyl	Vydate	–	–	–	–	X	–	–	2009
Permethrin	–	–	–	–	X	X	–	–	2009
Petroleum or paraffinic oil	Horticultural mineral oil	X	–	X	X	X	–	–	–
Phosmet	Imidan	X	–	X	–	X	–	–	2000–09
Pyrethrins/pyrethrum	–	–	–	–	X	–	–	–	–
Pyriproxyfen	–	X	–	–	–	X	–	–	2003–09
Rotenone	–	–	–	–	X	–	–	–	–
Rynaxypyr	–	–	–	–	–	X	–	–	–
Spinetoram	–	X	–	X	–	X	–	–	–
Spinosad	Entrust, Success	X	–	X	X	X	X	–	–
Spirodiclofen	Envidor	–	–	X	–	X	–	–	–
Spirotetramat	–	–	–	X	–	X	–	–	–
Tebufenozide	Confirm	X	–	–	–	–	–	–	–
Thiacloprid	–	–	–	–	–	X	–	–	–
Thiamethoxam	Axtara, Platinum, Voliam Flexi, Actara	X	–	X	–	X	–	–	–
Zeta cypermethrin	Mustang Max	X	–	–	–	–	X	–	–
Miticides									–
Acequinocyl	Kanemite	–	–	–	–	X	–	–	–
Bifenzate	Acramite	–	–	–	–	X	–	–	–
Clofentezine	Apollo	–	–	–	–	X	–	–	–
Etoxazole	Zeal	–	–	–	–	X	–	–	–
Fenpyroximate	Fujimite	–	–	–	–	X	–	–	–
Formetanate hydrochloride	Carzol	–	–	–	–	X	–	–	–
Hexythiazox	Onager, Savey	–	–	–	–	X	–	–	–
Propargite	Omite	–	–	–	–	X	–	–	–
Pyridaben	Nexter	–	–	–	–	X	–	–	–

[1] Also used as an insecticide.

Appendix D. U.S. Environmental Protection Agency and Oregon Water-Quality Criteria and U.S. Environmental Protection Agency Aquatic Life Benchmarks for Detected Pesticides

[**Source:** Oregon Department of Environmental Quality, 2004; U.S. Environmental Protection Agency, 2005a, 2009b. Concentrations in micrograms per liter; **Abbreviations:** USEPA, U.S. Environmental Protection Agency; CAS, Chemical Abstracts Service; CMC, criteria maximum concentration; CCC, criterion continuous concentration; –, no water-quality standard; <, less than; >, greater than]

| Pesticide | CAS number | USEPA Office of Pesticide Programs Aquatic Life Benchmarks | | | | USEPA Office of Water Aquatic Life Criteria | | Oregon Water Quality Criteria (freshwater) | |
| | | Fish | | Invertebrates | | Maximum Concentration (CMC) | Continuous Concentration (CCC) | | |
		Acute	Chronic	Acute	Chronic			Acute	Chronic
Atrazine	1912249	2,650	65	360	60	–	–	–	–
Azinphos-methyl	86500	0.18	0.055	0.08	0.036	–	–	–	0.01
Carbaryl	63252	110	6.8	0.85	0.5	–	–	–	–
Chlorpyrifos	2921882	0.9	0.57	0.05	0.04	0.083	0.041	0.083	0.041
DEET	134623	–	–	–	–	–	–	–	–
Diazinon	333415	45	< 0.55	0.105	0.17	0.17	0.17	–	–
Diuron	330541	355	26	80	160	–	–	–	–
Endrin	72208	–	–	–	–	0.086	0.036	0.18	0.0023
Fluometuron	2164172	320	–	110	–	–	–	–	–
Hexazinone	51235042	137,000	17,000	75,800	20,000	–	–	–	–
Imidacloprid	105827789	> 41,500	1,200	35	1.05	–	–	–	–
Malathion	121755	0.295	0.014	0.005	0.000026	–	0.1	–	0.1
Methomyl	16752775	160	12	2.5	0.7	–	–	–	–
Norflurazon	27314132	4,050	770	> 750	1,000	–	–	–	–
Phosmet	732116	35	3.2	1	0.8	–	–	–	–
Propiconazole	60207901	425	95	2,400	–	–	–	–	–
Propoxur	114261	1,850	–	5.5	–	–	–	–	–
Pyraclostrobin	175013180	–	–	–	–	–	–	–	–
Simazine	122349	3,200	960	500	2,000	–	–	–	–

Appendix E. Use and Environmental Fate Summaries for Detected Pesticides

Pesticide use, toxicity, and ecological transport and fate information provided below is from USEPA pesticide Reregistration Eligibility Decisions and other cited sources. USEPA toxicity classifications are assigned to fish and aquatic invertebrates based on reported acute toxicity (LC_{50} or EC_{50}) values and are summarized in table E1.

Atrazine (herbicide)

Atrazine is a triazine herbicide that targets grasses and broadleaf weeds (U.S. Environmental Protection Agency, 2006b). Statewide, it was the second most common pesticide applied for forestry in 2007 and 2008, although it is not commonly used in forests in the Hood River basin (Oregon Department of Agriculture, 2008, 2009; Doug Thiesies, Oregon Department of Forestry, oral commun., 2010). It is also approved for use on range grasses in Oregon under the U.S. Department of Agriculture's Conservation Reserve Program (U.S. Environmental Protection Agency, 2006b). It is mobile and persistent in the environment, with anaerobic half-lives of 330 and 578 days in water and sediment, respectively (U.S. Environmental Protection Agency, 2006b). Aerobic half-lives are estimated as 30 and 87–146 days in water and soil, respectively (U.S. Environmental Protection Agency, 2006b; California Environmental Protection Agency, 2010). In water and soils, it breaks down more slowly under neutral than acidic or basic conditions (U.S. Environmental Protection Agency, 2006b). Microbial metabolism is the main degradation pathway in aerobic environments (U.S. Environmental Protection Agency, 2006b). Atrazine easily washes off of foliage and commonly enters surface waters during the first precipitation event following application (U.S. Environmental Protection Agency, 2006b). Because it does not sorb strongly to soils, it is highly mobile in the environment and has potential to contaminate groundwater (U.S. Environmental Protection Agency, 2006b; California Environmental Protection Agency, 2010). Due to its potential to contaminate surface and groundwaters, atrazine is a Restricted Use Pesticide. It is moderately toxic to fish and moderately to highly toxic to freshwater invertebrates (U.S. Environmental Protection Agency, 2006b). Reductions in fish populations are estimated to occur at 62 µg/L (species not specified) (U.S. Environmental Protection Agency, 2006b).

Table E1. U.S. Environmental Protection Agency toxicity classifications for fish and aquatic invertebrates.

[**Source:** U.S. Environmental Protection Agency, 2010b. **Abbreviations:** LC_{50}, 50 percent lethal concentration; EC_{50}, 50 percent effective concentration; mg/L, milligrams per liter; <, less than; >, greater than]

Toxicity Category	LC_{50} or EC_{50} (mg/L)
Very highly toxic	<0.1
Highly toxic	0.1–1
Moderately toxic	>1 <10
Slightly toxic	>10 <100
Practically nontoxic	>100

Azinphos-methyl (insecticide)

Azinphos-methyl is an organophosphate insecticide used on fruit, nut, and vegetable crops, with no residential or public health uses (U.S. Environmental Protection Agency, 2006a). Historically, it has been widely used on orchards in the Hood River basin, but its use on all crops, including tree fruits and blueberries, will be phased out by September 30, 2012 (U.S. Environmental Protection Agency, 2009c; Steve Castagnoli, Oregon State University Extension Service, oral commun., 2010). Major pathways to surface waters are spray drift, runoff, and foliar wash-off (U.S. Environmental Protection Agency, 2006a). Azinphos-methyl is relatively insoluble in water, mobile and moderately persistent in soils, and like other organophosphates, degrades relatively quickly in water (U.S. Environmental Protection Agency, 2006a; Oregon State University, 1996a). Field studies indicate that degradation products are less toxic than the parent product (U.S. Environmental Protection Agency, 2006a). It is very highly toxic to freshwater fish and invertebrates and is a Toxicity Category I (highly toxic) pesticide, labeled as a Restricted Use Product (U.S. Environmental Protection Agency, 2006a).

Carbaryl (insecticide)

Carbaryl is a broad-spectrum carbamate insecticide. Orchard uses are primarily in apples and cherries. It is also a fruit thinning agent for apples. It is the active ingredient in several pesticide products registered for nonagricultural uses including turf, ornamental, and residential use. Carbaryl is moderately soluble in water and does not bind strongly to soils (it is mobile), but its adsorption tendency depends on the soil organic matter content (National Marine Fisheries Service, 2009). Half-lives for microbial metabolism are 4–5 days in aerobic soil and water (National Marine Fisheries Service, 2009). Carbaryl hydrolyzes quickly, with half-lives ranging from 3.2 hours to 12 days at basic and neutral pH (9 and 7, respectively) (National Marine Fisheries Service, 2009). The mean pH ranged from 7.1 to 8.0 in Hood River basin streams during the months when carbaryl was detected in those streams in 2009, so half-lives of carbaryl in the basin are expected to be in that range. In rivers of that pH range, it has been shown to degrade completely within 2 weeks (National Pesticide Information Center, 2003). Carbaryl's persistence in the environment is prolonged under acidic or anaerobic conditions (National Marine Fisheries Service, 2009). Its major degradation product is 1-napthanol, which further degrades to carbon dioxide (National Marine Fisheries Service, 2009). Carbaryl can be very highly toxic to fish (rainbow trout) and aquatic invertebrates. The USEPA and National Marine Fisheries Service are looking into its effects on endangered salmon (National Marine Fisheries Service, 2009). The major degradation product, 1-napthanol, ranges from moderately to highly toxic to aquatic organisms (National Marine Fisheries Service, 2009). As of May 2010, the USEPA is implementing new restrictions on the use of carbaryl and other n-methyl carbamate pesticides to protect threatened or endangered Pacific salmon and steelhead (U.S. Environmental Protection Agency, 2010a).

Chlorpyrifos (insecticide)

In the last decade, chlorpyrifos was one of the most commonly used organophosphate insecticides in the United States, with applications in food crops, cattle ear tags, containerized baits, wood treatments, golf courses, Christmas trees, and public health (mosquito and fire ant control). Its use has been phased out for structural control of termites and most residential applications (U.S. Environmental Protection Agency, 2006c). It can be used in the late winter and early spring on pears, cherries, and apples in the Hood River basin, with use patterns fluctuating annually (Steve Castagnoli, Oregon State University Extension Service, oral commun., 2010; Oregon State University Extension Service, 2010). The USEPA is currently planning to implement new restrictions on the use of chlorpyrifos near salmon-bearing streams (U.S. Environmental Protection Agency, 2010a). Chlorpyrifos has low mobility, strong sediment binding capacity, and low water

solubility (U.S. Environmental Protection Agency, 2006c). It is moderately persistent in soils (U.S. Environmental Protection Agency, 2006c). Spray drift during application and adsorption to eroding soil are major pathways into surface waters (U.S. Environmental Protection Agency, 2006c). Its persistence in water depends on the formulation and environmental conditions, with faster degradation rates with increasing temperature and pH (Oregon State University, 1996b). Volatilization seems to be the main pathway of loss from water (Oregon State University, 1996b). Its major degradation product, TCP (3,5,6-trichloropyridinol), is more mobile and persistent than chlorpyrifos, making it more likely to be found in the dissolved phase and available for aqueous runoff to streams (U.S. Environmental Protection Agency, 2006c). TCP was not analyzed for this project.

Chlorpyrifos is highly toxic to fish and very highly toxic to aquatic invertebrates (U.S. Environmental Protection Agency, 2006c). Because of its toxicity to prey items for threatened and endangered Pacific salmon and steelhead, the USEPA expressed "significant concern" over considerable use of chlorpyrifos where it can enter salmonid habitats (National Marine Fisheries Service, 2008). In salmon-bearing basins, reduced salmonid prey availability has been correlated to organophosphate use (National Marine Fisheries Service, 2008). Sublethal concentrations of chlorpyrifos have been shown to inhibit swimming and olfactory-mediated behaviors in salmonids (National Marine Fisheries Service, 2008; Tierney and others, 2010). Containers less than 15 gallons (liquid) or 25 pounds (dry) and all emulsifiable concentrate end-use products are labeled as Restricted Use Pesticides (U.S. Environmental Protection Agency, 2006c).

DEET (insecticide)

DEET is an insect and acarid (mite and tick) repellent in the N,N-dialkylamide chemical family (National Pesticide Information Center, 2008). It is approved for use in households, on the human body, and on pets (U.S. Environmental Protection Agency, 1998a). It is moderately mobile in soils and is stable to hydrolysis in soils at common pH ranges (National Pesticide Information Center, 2008). DEET is practically insoluble in water and has been detected in streams that receive wastewater, as most DEET absorbed through the skin is excreted through urine (U.S. Environmental Protection Agency, 1998a; National Pesticide Information Center, 2008). DEET is slightly toxic to freshwater fish (rainbow trout) and invertebrates (U.S. Environmental Protection Agency, 1998a).

Diazinon (insecticide)

Diazinon has been one of the most commonly used organophosphate insecticides in the United States for agricultural and household uses, although residential uses were phased out in 2004 (U.S. Environmental Protection Agency, 2006d, 2008). It can be used in the spring and summer in the Hood River basin on orchard crops and blueberries (Hollingsworth, 2009; Oregon State University Extension Service, 2010). It is moderately persistent in soils. Its persistence in surface waters varies with pH; its hydrolysis half-life is an order of magnitude higher (more slowly degrading) at pH 7 than at pH 5 (U.S. Environmental Protection Agency, 2006d). Due to diazinon's mobility, runoff is a common pathway to surface waters. Diazinon is very highly toxic to fish and even more so to aquatic invertebrates after acute or chronic exposures (U.S. Environmental Protection Agency, 2006d). Diazinon has also been shown to increase vulnerability to predation (U.S. Environmental Protection Agency, 2006d) and impair swimming, olfaction, and olfactory-mediated behaviors of salmonids at sublethal concentrations (National Marine Fisheries Service, 2008). Because of diazinon's potential to harm salmonids, the USEPA is implementing new restrictions on the use of diazinon near streams (National Marine Fisheries Service, 2008; U.S. Environmental Protection Agency, 2010a).

Diuron (herbicide)

Diuron is a substituted urea herbicide that is used to control emerging and young broadleaf weeds, grasses, and mosses on a variety of agricultural crops and in dry irrigation canals, and to control algae in ponds (U.S. Environmental Protection Agency, 2003). It is commonly used in the early spring through early summer in the Hood River basin for orchard and rights-of-way weed control (Steve Castagnoli, Oregon State University Extension Service, oral commun., 2010; Brian Walker, Oregon Department of Transportation, oral commun., 2010). Diuron has potential to contaminate groundwater and is persistent in soils, with soil half-lives ranging from 1 month to 1 year (Oregon State University, 1996c; U.S. Environmental Protection Agency, 2003; Peachey, 2009). It is generally stable in surface waters, with microbial degradation as the main mode of loss (Oregon State University, 1996c; U.S. Environmental Protection Agency, 2003). Toxicity to salmonids ranges from slightly toxic (coho salmon) to highly toxic (Chinook salmon and cutthroat trout) (U.S. Environmental Protection Agency, 2003, 2007). Diuron is moderately to highly toxic to aquatic invertebrates (U.S. Environmental Protection Agency, 2003).

Endrin (insecticide)

Endrin is an organochlorine insecticide, rodenticide, and avicide for which all uses have been cancelled in the United States since 1991. Among other uses, it was used to control rodents in orchards (U.S. Environmental Protection Agency, 2009a). It binds strongly to soils and has been shown to persist for 14 years or more (Agency for Toxic Substances and Disease Registry, 1996; California Environmental Protection Agency, 1999; U.S. Environmental Protection Agency, 2009a). It can reach surface waters through erosion of contaminated soils (U.S. Environmental Protection Agency, 2009a). Because endrin is very insoluble in water, detections in surface water are rare (California Environmental Protection Agency, 1999). Endrin has been associated with fish and bird kills. It is very highly toxic to fish, including salmonids, and highly to very highly toxic to aquatic invertebrates (U.S. Environmental Protection Agency, 2007).

Fluometuron (herbicide)

Fluometuron is a substituted urea herbicide used to control broadleaf weeds and annual grasses. It is currently only approved for use on cotton, although it was registered for use on sugarcane until 1998 (U.S. Environmental Protection Agency, 2005e). No Federal Insecticide, Fungicide, and Rodenticide Act (FIFRA) Section 24(c) Special Local Needs permits for fluometuron are registered in Oregon, so its presence in the basin is unexpected. It is mobile and persistent in soils, with an aerobic soil half-life of 181 days (U.S. Environmental Protection Agency, 2005c). It is stable to hydrolysis and photolysis, so it also persists in surface and groundwaters (U.S. Environmental Protection Agency, 2005e). Based on its toxicity to rainbow trout (*Oncorhynchus mykiss*), fluometuron is moderately toxic to freshwater fish, although it ranges from slightly to highly toxic to non-salmonid fish species (U.S. Environmental Protection Agency, 2005c).

Hexazinone (herbicide)

Hexazinone is a triazine herbicide registered for use in forestry, pasture and rangeland, rights-of-way, and on blueberries (U.S. Environmental Protection Agency, 1994; Peachey, 2009). In the Hood River basin, it is commonly used in forests (Doug Thiesies, Oregon Department of Forestry, oral commun., 2010). It has high potential to contaminate surface- and groundwaters from spray drift, runoff (even months after its application), and leaching (U.S. Environmental Protection Agency, 1994). It requires rainfall or irrigation for activation and is mobile due to its high solubility in water and low soil adsorption tendency (U.S. Environmental Protection Agency, 1994; Oregon State University, 1996d; U.S. Department of Agriculture, 2009). It is also persistent in soil and water. Microbial degradation is the primary breakdown mechanism; photodegradation and hydrolysis are relatively unimportant (U.S. Environmental Protection Agency, 1994). Half-lives in nonsterile aerobic soils have been shown to range from 216 to 228 days and can be greater than 2 months in nonsterile aerobic waters (U.S. Environmental Protection Agency, 1994). Hexazinone is practically nontoxic in acute exposures to rainbow trout (LC_{50} >320 mg/L) (U.S. Environmental Protection Agency, 1994). Sublethal effects of hexazinone exposure on salmonids are unknown. It is practically nontoxic to slightly toxic to aquatic invertebrates (U.S. Environmental Protection Agency, 1994; Oregon State University, 1996d).

Imidacloprid (insecticide)

Imidacloprid is a neonicotinoid (chloro-nicotinyl) insecticide for use on structures, orchards, grapes, blueberries, and soil and seeds, and for residential use to control sucking insects and some chewing insects (National Pesticide Information Center, 2010; Oregon State University Extension Service, 2010). Neonicotinoid insecticides have become common replacements for some organophosphate insecticides, many of which are being phased out by the USEPA due to their toxicity. Soil half-lives for imidacloprid range from 40 days on unamended soil to 124 days on soil enriched with organic fertilizer (National Pesticide Information Center, 2010). Soil sorption of imidacloprid increases with increasing soil organic matter content, imidacloprid concentration, and time (National Pesticide Information Center, 2010). Its moderate soil binding capacity and water solubility lend it to being more mobile in porous, gravelly, or cobbly soils (Oregon State University, 1996e). In water, imidacloprid breaks down primarily by photolysis, although hydrolysis occurs increasingly with increasing pH and temperature (National Pesticide Information Center, 2010). Imidacloprid is a general use pesticide that is slightly toxic to rainbow trout, but highly to very highly toxic to aquatic invertebrates (Oregon State University, 1996e; National Pesticide Information Center, 2010).

Malathion (insecticide)

Malathion is a broad-range organophosphate insecticide and miticide with agricultural, industrial, public health, and residential applications (U.S. Environmental Protection Agency, 2009d). It is registered for use on orchards, grapes, blueberries, rangeland and hay, and in residences, and has been used to control the western cherry fruit fly in the Hood River basin (Hollingsworth, 2009; Joe McCanna, the Confederated Tribes of Warm Springs, oral commun., 2010).

Malathion is relatively nonpersistent in the environment, with aerobic soil half-lives on the order of hours to days, decreasing with increasing soil moisture, microbial activity, or pH (U.S. Environmental Protection Agency, 2009d). Half-lives in surface water range from 1 day to 2 weeks, but malathion has potential to contaminate groundwater due to its solubility and moderate soil-adsorption factor, although it is expected to have low persistence in anaerobic waters (Oregon State University, 1996f; U.S. Environmental Protection Agency, 2009d). Malathion reaches streams through off-target drift, agricultural runoff, and urban runoff from residential and public health or quarantine uses over broad areas (U.S. Environmental Protection Agency, 2009d). Malathion is a general use pesticide that is highly toxic to fish and aquatic invertebrates (Oregon State University, 1996f; U.S. Environmental Protection Agency, 2009d). Sublethal concentrations of malathion can impair swimming and reproduction or growth of salmonids and survival of their prey (National Marine Fisheries Service, 2008). Effects on olfaction and olfactory-mediated behaviors in salmonids exposed to malathion have not been assessed (National Marine Fisheries Service, 2008). The USEPA is imposing new limitations on the use of malathion near salmon-bearing streams because of its potential to harm salmonids (National Marine Fisheries Service, 2008; U.S. Environmental Protection Agency, 2010a).

Methomyl (insecticide)

Methomyl is a carbamate insecticide registered for control of a broad range of insect pests on a wide variety of food (including orchard) and feed crops, in livestock feedlots and sleeping quarters, food processing facilities, and other commercial settings (U.S. Environmental Protection Agency, 1998b). It is not known to be in common use in the Hood River basin (Steve Castagnoli, Oregon State University Extension Service, written commun., 2010). It can reach surface waters via runoff, erosion, or spray drift and can leach to groundwater (U.S. Environmental Protection Agency, 1998b; National Marine Fisheries Service, 2009). Because it is highly mobile and moderately persistent in the environment, it can be available for runoff for days to weeks following application (U.S. Environmental Protection Agency, 1998b). It is not expected to persist in shallow, clear waters due to its susceptibility to photolysis, but lasts relatively longer in aerobic soils, with half-lives ranging from 11 to 45 days (U.S. Environmental Protection Agency, 1998b; National Marine Fisheries Service, 2009). Methomyl is moderately to highly toxic to freshwater fish and highly to very highly toxic to aquatic invertebrates (U.S. Environmental Protection Agency, 1998b). The USEPA has designated it as a Restricted Use Pesticide because of its toxicity to humans (Oregon State University, 1996g).

Norflurazon (herbicide)

Norflurazon is a selective herbicide in the fluorinated pyridazinone chemical class that targets broadleaf weeds, grasses, and sedges (U.S. Environmental Protection Agency, 1996). It is used on a variety of food crops, including pears, apples, and cherries, on rights-of-ways, uncultivated agricultural and nonagricultural areas, and in outdoor industrial areas (U.S. Environmental Protection Agency, 1996). Norflurazon can contaminate surface waters via spray drift and runoff, which can occur several months after application (U.S. Environmental Protection Agency, 1996). In surface waters, the primary mode of loss is photodegradation, which has a half-life of less than 3 days (U.S. Environmental Protection Agency, 1996). In soils, it breaks down more slowly, with half-lives of 130 days and 6–8 months under aerobic and anaerobic conditions, respectively. It is mobile to highly mobile in soil (U.S. Environmental Protection Agency, 1996). Norflurazon is moderately toxic to rainbow trout and slightly toxic to freshwater invertebrates (U.S. Environmental Protection Agency, 1996).

Phosmet (insecticide)

Phosmet is a broad-range organophosphate insecticide with various agricultural and forestry uses. Household use for ornamental or tree fruits and dogs was cancelled in 2001 (U.S. Environmental Protection Agency, 2001). Use is still allowed on apples, cherries, blueberries, grapes, evergreen trees, pears, and many other crops (U.S. Environmental Protection Agency, 2001). Phosmet binds moderately to soils and has low water solubility so is not highly mobile (Oregon State University, 1996h; U.S. Environmental Protection Agency, 2006e). It degrades rapidly in aerobic soils, chiefly due to microbial degradation and hydrolysis (Oregon State University, 1996h; U.S. Environmental Protection Agency, 2006e). Phosmet can reach surface waters from drift due to aerial or ground surface spray (U.S. Environmental Protection Agency, 2006e). Dissolved-phase phosmet can contaminate surface waters through runoff if precipitation or irrigation occur soon after its application (U.S. Environmental Protection Agency, 2006e). However, it does not persist in water; it degrades by hydrolysis and photolysis, with half-lives ranging from 16 hours at pH 9 to 9 days at pH 5 (Oregon State University, 1996h). Phosmet toxicity to salmonids ranges from moderately toxic (rainbow trout) to highly toxic (Chinook salmon) (U.S. Environmental Protection Agency, 2007). It ranges from moderate to very highly toxic for macroinvertebrates and zooplankton (U.S. Environmental Protection Agency, 2007).

Propiconazole (fungicide)

Propiconazole is a broad-range systemic foliar fungicide used to control powdery mildews, rust, and leaf spots on cherries and household crops (U.S. Environmental Protection Agency, 2006f). It is highly persistent in soil, but photodegrades rapidly in water (U.S. Environmental Protection Agency, 2006f). It is relatively immobile in highly organic soils and moderately mobile in soils low in organic matter (U.S. Environmental Protection Agency, 2006f). It is considered very highly toxic to freshwater fish.

Propoxur (insecticide)

Propoxur is a carbamate insecticide that targets ants, bees, cockroaches, fleas, mosquitoes, spiders, and wasps for use in homes, on pets, and on pavement and commercial structures (U.S. Environmental Protection Agency, 1997). Because it is mostly used indoors, the USEPA requires less rigorous testing of environmental fate and transport compared to pesticides that are primarily used outside (U.S. Environmental Protection Agency, 1997). Although limited environmental fate data exist, some conclusions can be made. It is highly mobile and has transport characteristics similar to other pesticides that are known to leach to groundwater (U.S. Environmental Protection Agency, 1997). It is expected to be moderately persistent in soils, with a half-life of several months (U.S. Environmental Protection Agency, 1997). In water, its expected half-life is 13 days (U.S. Environmental Protection Agency, 1997). It is more stable under acidic and neutral than alkaline conditions (U.S. Environmental Protection Agency, 1997). Propoxur is moderately toxic to fish, but very highly toxic to aquatic invertebrates (U.S. Environmental Protection Agency, 1997). It is a General Use Pesticide.

Pyraclostrobin (fungicide)

Pyraclostrobin is a strobilurin fungicide used to kill blights, mildews, molds, and rusts and is registered for use on many crops, including cherries and several berry crops. It is moderately persistent in aerobic soils (half-life = 136 days), but less persistent in anaerobic soils (3 days) (California Environmental Protection Agency, 2010). In aerobic water, its half-life ranges from 1 to 4 days (New York State Department of Environmental Conservation, 2004). Its high organic-carbon partitioning coefficient (K_{oc}) and low solubility in water indicate that it will strongly bind to soil organic matter, so it is largely immobile in soils (California Environmental Protection Agency, 2010). Although few ecotoxicological data exist for pyraclostrobin, it is considered very highly toxic to aquatic organisms, particularly rainbow trout (BASF Corporation, 2010).

Simazine (herbicide)

Simazine is a selective triazine herbicide used to control annual grasses and broadleaf weeds, usually applied to the soil either before emergence or after removal of weed growth (U.S. Environmental Protection Agency, 2006g). It is widely used in Hood River basin orchards and is also registered for use on blueberries and vineyards (Peachey, 2009; Steve Castagnoli, Oregon State University Extension Service, oral commun., 2010). Its main pathways to surface waters on the Pacific coast are spray drift and runoff (U.S. Environmental Protection Agency, 2006g). Simazine is highly mobile, particularly in soils with low organic matter content, where its potential for groundwater contamination is high (U.S. Environmental Protection Agency, 2006g). It is also persistent in the environment, with half-lives in soil ranging from 22 to 664 days, depending on sunlight and oxygen availability, and aqueous half-lives ranging from 12 to 700 days (U.S. Environmental Protection Agency, 2006g). Simazine is practically nontoxic to salmonids and slightly to moderately toxic to aquatic invertebrates (Oregon State University, 1996i; U.S. Environmental Protection Agency, 2007). Sublethal concentrations of simazine have been shown to impact reproduction and olfaction in Atlantic salmon (*Salmo salar*), swimming in rainbow trout (*O. mykiss*), and reduce the survival of various invertebrates that are potential salmonid prey items (U.S. Environmental Protection Agency, 2007; Tierney and others, 2010).

Appendix F.   Pesticides Analyzed in Samples Collected in the Hood River Basin, Oregon, 1999–2009

[Oregon Department of Environmental Quality method numbers are in parentheses. Detected pesticides are shown in bold print. Pesticides analyzed before 2009 are shown in italicized print.]

2,4-D (SM 6640)
4,4'-DDD (p,p'-DDD) (8270C)
4,4'-DDE (8270C)
4,4'-DDT (8270C)
Alachlor (8321)
Aldrin (8270C)
alpha-BHC (benzene hexachloride) (8270C)
Ametryn (8270C)
Aminocarb (8321)
Atraton (8270C)
Atrazine (8270, 8141B)
Azinphos-methyl (8270C)
Azinphos-methyl oxon (8141A)
bentazon (bentazone) (6640B)
beta-BHC (benzene hexachloride) (8270C)
Bromacil (8270C)
Butachlor (8270C)
Butylate (8270C)
Carbaryl (8321B)
Carbofuran (8321B)
Carboxin (8270C)
Chlorobenzilate (a) (8270C)
Chloroneb (8270C)
Chlorpyrifos (8321B, 8270C, 8141B)
Chlorpyrifos oxon (8141A)
Chlorothalonil (8270C)
Chlorpropham (8270C)
Cis-Chordane (8270C)
Cyanazine (8270C)
Cycloate (8270C)
Dacthal (DCPA) (Chlorthal-Dimethyl) (8270C)
DEET/N,N-Diethyl-meta-toluamide (8321, 8270C)
delta-BHC (benzene hexachloride) (8270C)
Diazinon (8321B)
Dichlorvos (8270C)
Dieldrin (8270C)
Dimethoate (8270C)
Diphenamid (8270C)
Disulfoton (8270C)
Diuron (8321)
Endosulfan 1 (8270C)
Endosulfan 2 (8270C)
Endosulfan Sulfate (8270C)
Endrin Aldehyde (8270C)
Endrin (8270C)
EPTC (Eptam) (8270C)
Ethoprophos (Ethoprop) (Prophos) (8270C)
Etridiazole (8270C)
Fenamiphos (8270C)
Fenarimol (8270C)
Fenvalerate + Esfenvalerate (8270C)
Fluometuron (8321)
Fluridone (8270C)
Heptachlor (8270C)
Hexachlorobenzene (8270C)

Hexachlorocyclopentadiene (8270C)
Hexazinone (8270) (8270C)
Imazapyr (8321)
Imidacloprid (8321B)
Lindane (gamma-BHC) (benzene hexachloride) (8270C)
Linuron (8321)
Malathion (8270) (8270C)
Malathion oxon (8141A)
Methiocarb (8321)
Methomyl (8321)
Methoxychlor (8270C)
Methyl paraoxon (8270C)
Methyl parathion (8270C)
Metolachlor (8270, 8270C, 83213)
Metribuzin (8270C)
Metribuzin (8321)
Mexacarbate (8321)
MGK-264 (N-octyl bicycloheptane dicarboximide) (8270C)
Molinate (8270C)
Napropamide (8270C)
Neburon (8321)
Norflurazon (8270C)
Oxyamyl (8321)
Pebulate (8270C)
Pendimethalin (8270C)
Pentachlorophenol (8270C)
Permethrin (8270)
Phosmet (8270C)
Phosmet oxon (8141A)
Prometon (8270C, 8321)
Prometryn (8270C, 8321)
Propoxur (8321)
Pronamide (Propyzamide) (8270C)
Propachlor (8270C)
Propazine (8270C, 8321)
Propiconazole (8321)
Pyraclostrobin (8321)
Pyriproxyfen (8270C)
Siduron (8321)
Simazine (8270, 8270C, 8321)
Simetryn (8270C, 8321)
Tebuthiuron (8270C)
Terbacil (8270C)
Terbufos (8270C)
Terbutryne (8270C, 8321)
Terbutylazine (8321)
Tetrachlorvinphos (8270C)
Trans-Chordane (8270C)
Trans-Nonachlor (8270C)
Triadimefon (8270C)
Triclopyr (8321) (6640B)
Tricyclazole (8270C)
Trifluralin (8270C)
Vernolate (8270C)

Appendix G. Sample and Detection Counts by Pesticide in Samples Collected from the Hood River Basin, Oregon, 1999–2009

Data in the following tables were screened to the reporting limit indicated in the caption. Data were screened so that differences in reporting limits from year to year would not skew the assessment of detection trends (refer to the Methods section for more information on data screening).

Table G1. Atrazine sample and detection counts, Hood River basin, Oregon, 1999–2009, using data screened at 0.027 micrograms per liter.

[**Abbreviations:** d, number of samples with detections at the screening level (0.027 micrograms per liter); n, number of samples; RM, river mile; –, not sampled]

Atrazine	1999		2000		2001		2002		2003		2004	
	d	n	d	n	d	n	d	n	d	n	d	n
Baldwin	–	–	–	–	–	–	–	–	0	19	0	12
Dog	–	–	–	–	0	11	0	7	0	11	0	12
Evans	–	–	–	–	0	14	0	15	0	20	0	12
Hood, East Fork	0	1	0	12	–	–	0	15	–	–	–	–
Hood, Middle Fork	0	1	0	11	–	–	–	–	–	–	–	–
Hood, mouth	0	3	0	2	0	11	0	7	–	–	–	–
Hood, West Fork, mouth	–	–	–	–	–	–	–	–	–	–	–	–
Hood, West Fork, RM 2.5	–	–	–	–	–	–	–	–	–	–	–	–
Hood, West Fork, RM 4.7	0	3	0	11	0	1	–	–	–	–	–	–
Indian	0	2	–	–	–	–	–	–	–	–	–	–
Lenz	–	–	–	–	0	14	1	15	1	20	0	17
Neal, middle	–	–	–	–	–	–	–	–	–	–	0	9
Neal, mouth	0	3	0	12	0	16	0	17	0	21	0	17
Neal, upper, above diversion	–	–	–	–	0	17	0	20	0	21	0	17
Neal, upper, below diversion	–	–	–	–	–	–	–	–	0	21	0	11
Rogers	–	–	–	–	–	–	–	–	–	–	–	–
Total	0	13	0	48	0	84	1	96	1	133	0	107

Atrazine	2005		2006		2007		2008		2009		Total	
	d	n	d	n	d	n	d	n	d	n	d	n
Baldwin	0	18	0	3	–	–	–	–	–	–	0	52
Dog	–	–	–	–	–	–	–	–	–	–	0	41
Evans	0	19	0	3	–	–	–	–	–	–	0	83
Hood, East Fork	–	–	–	–	–	–	–	–	–	–	0	28
Hood, Middle Fork	–	–	–	–	–	–	–	–	–	–	0	12
Hood, mouth	0	15	0	7	0	13	0	10	0	12	0	80
Hood, West Fork, mouth	–	–	–	–	–	–	0	5	0	12	0	17
Hood, West Fork, RM 2.5	–	–	–	–	–	–	0	5	0	1	0	6
Hood, West Fork, RM 4.7	–	–	–	–	–	–	–	–	–	–	0	15
Indian	–	–	–	–	–	–	–	–	–	–	0	2
Lenz	0	17	0	3	–	–	0	7	0	12	2	105
Neal, middle	0	16	0	6	0	13	0	10	0	13	0	67
Neal, mouth	0	17	0	7	0	13	0	10	0	12	0	145
Neal, upper, above diversion	0	19	0	7	0	12	–	–	–	–	0	113
Neal, upper, below diversion	0	16	0	7	0	13	0	10	0	13	0	91
Rogers	–	–	–	–	–	–	0	10	0	12	0	22
Total	0	137	0	43	0	64	0	67	0	87	2	879

Table G2.   Azinphos-methyl sample and detection counts, Hood River basin, Oregon, 1999–2009, using data screened at 0.03 micrograms per liter.

[**Abbreviations:** d, number of samples with detections at the reporting level (0.03 micrograms per liter); n, number of samples; RM, river mile; –, not sampled]

Azinphos–methyl	1999		2000		2001		2002		2003		2004	
	d	n	d	n	d	n	d	n	d	n	d	n
Baldwin	–	–	–	–	–	–	–	–	0	19	0	12
Dog	–	–	–	–	0	11	0	7	0	11	0	12
Evans	–	–	–	–	2	14	0	15	0	20	0	12
Hood, East Fork	0	1	0	12	–	–	0	15	–	–	–	–
Hood, Middle Fork	0	1	0	12	–	–	–	–	–	–	–	–
Hood, mouth	0	1	0	2	0	11	0	7	–	–	–	–
Hood, West Fork, mouth	–	–	–	–	–	–	–	–	–	–	–	–
Hood, West Fork, RM 2.5	–	–	–	–	–	–	–	–	–	–	–	–
Hood, West Fork, RM 4.7	0	1	0	11	0	1	–	–	–	–	–	–
Indian	–	–	–	–	–	–	–	–	–	–	–	–
Lenz	–	–	–	–	0	17	3	15	14	20	5	17
Neal, middle	–	–	–	–	–	–	–	–	–	–	0	9
Neal, mouth	1	1	1	15	1	16	4	17	11	21	2	17
Neal, upper, above diversion	–	–	–	–	0	17	0	20	0	21	0	17
Neal, upper, below diversion	–	–	–	–	–	–	–	–	0	21	0	11
Rogers	–	–	–	–	–	–	–	–	–	–	–	–
Total	1	5	1	52	3	87	7	96	25	133	7	107

Azinphos–methyl	2005		2006		2007		2008		2009		Total	
	d	n	d	n	d	n	d	n	d	n	d	n
Baldwin	0	18	0	3	–	–	–	–	–	–	0	52
Dog	–	–	–	–	–	–	–	–	–	–	0	41
Evans	0	19	0	3	–	–	–	–	–	–	2	83
Hood, East Fork	–	–	–	–	–	–	–	–	–	–	0	28
Hood, Middle Fork	–	–	–	–	–	–	–	–	–	–	0	13
Hood, mouth	0	15	0	7	0	13	0	9	0	15	0	80
Hood, West Fork, mouth	–	–	–	–	–	–	0	5	0	14	0	19
Hood, West Fork, RM 2.5	–	–	–	–	–	–	0	4	0	1	0	5
Hood, West Fork, RM 4.7	–	–	–	–	–	–	–	–	–	–	0	13
Indian	–	–	–	–	–	–	–	–	–	–	–	–
Lenz	6	18	1	3	–	–	0	6	0	15	29	111
Neal, middle	2	16	0	7	0	13	0	9	0	15	2	69
Neal, mouth	3	17	4	7	1	13	0	9	0	15	28	148
Neal, upper, above diversion	1	19	0	7	0	12	–	–	–	–	1	113
Neal, upper, below diversion	0	16	0	7	0	13	0	9	0	15	0	92
Rogers	–	–	–	–	–	–	0	9	0	14	0	23
Total	12	138	5	44	1	64	0	60	0	104	62	890

Table G3. Chlorpyrifos sample and detection counts, Hood River basin, Oregon, 1999–2009, using data screened at 0.03 micrograms per liter.

[**Abbreviations:** d, number of samples with detections at the reporting level (0.03 micrograms per liter); n, number of samples; RM, river mile; –, not sampled]

Chlorpyrifos	1999		2000		2001		2002		2003		2004	
	d	n	d	n	d	n	d	n	d	n	d	n
Baldwin	–	–	–	–	–	–	–	–	0	19	0	12
Dog	–	–	–	–	0	11	0	7	0	11	0	12
Evans	–	–	–	–	1	14	0	15	0	20	0	12
Hood, East Fork	0	1	0	12	–	–	0	15	–	–	–	–
Hood, Middle Fork	0	1	0	11	–	–	–	–	–	–	–	–
Hood, mouth	1	3	0	2	1	11	0	7	–	–	–	–
Hood, West Fork, mouth	–	–	–	–	–	–	–	–	–	–	–	–
Hood, West Fork, RM 2.5	–	–	–	–	–	–	–	–	–	–	–	–
Hood, West Fork, RM 4.7	0	3	0	11	0	1	–	–	–	–	–	–
Indian	2	2	–	–	–	–	–	–	–	–	–	–
Lenz	–	–	–	–	4	17	2	15	2	20	1	17
Neal, middle	–	–	–	–	–	–	–	–	–	–	0	9
Neal, mouth	2	3	5	12	5	16	3	17	1	21	1	17
Neal, upper, above diversion	–	–	–	–	0	17	0	20	0	21	0	17
Neal, upper, below diversion	–	–	–	–	–	–	–	–	0	21	0	11
Rogers	–	–	–	–	–	–	–	–	–	–	–	–
Total	5	13	5	48	11	87	5	96	3	133	2	107

Chlorpyrifos	2005		2006		2007		2008		2009		Total	
	d	n	d	n	d	n	d	n	d	n	d	n
Baldwin	0	18	0	3	–	–	–	–	–	–	0	52
Dog	–	–	–	–	–	–	–	–	–	–	0	41
Evans	1	19	0	3	–	–	–	–	–	–	2	83
Hood, East Fork	–	–	–	–	–	–	–	–	–	–	0	28
Hood, Middle Fork	–	–	–	–	–	–	–	–	–	–	0	12
Hood, mouth	0	15	0	7	0	13	0	10	1	11	2	68
Hood, West Fork, mouth	–	–	–	–	–	–	0	5	0	9	0	5
Hood, West Fork, RM 2.5	–	–	–	–	–	–	0	5	0	1	0	5
Hood, West Fork, RM 4.7	–	–	–	–	–	–	–	–	–	–	0	15
Indian	–	–	–	–	–	–	–	–	–	–	2	2
Lenz	2	18	0	3	–	–	0	7	3	13	11	97
Neal, middle	2	16	0	7	0	13	0	10	1	13	2	55
Neal, mouth	3	17	0	7	0	13	0	10	4	13	20	133
Neal, upper, above diversion	0	19	0	7	0	12	–	–	–	–	0	113
Neal, upper, below diversion	0	16	0	7	0	13	1	10	0	14	1	78
Rogers	–	–	–	–	–	–	0	10	0	14	0	10
Total	8	138	0	44	0	64	1	67	9	88	40	797

Table G4. Diazinon sample and detection counts, Hood River basin, Oregon, 1999–2009, using data screened at 0.1 micrograms per liter.

[**Abbreviations:** d, number of samples with detections at the reporting level (0.1 micrograms per liter); n, number of samples; RM, river mile; –, not sampled]

Diazinon	1999		2000		2001		2002		2003		2004	
	d	n	d	n	d	n	d	n	d	n	d	n
Baldwin	–	–	–	–	–	–	–	–	0	19	0	12
Dog	–	–	–	–	0	11	0	7	0	11	0	12
Evans	–	–	–	–	0	14	0	15	0	20	0	12
Hood, East Fork	0	1	0	16	–	–	0	15	–	–	–	–
Hood, Middle Fork	0	1	0	15	–	–	–	–	–	–	–	–
Hood, mouth	0	4	0	3	0	11	0	7	–	–	–	–
Hood, West Fork, mouth	–	–	–	–	–	–	–	–	–	–	–	–
Hood, West Fork, RM 2.5	–	–	–	–	–	–	–	–	–	–	–	–
Hood, West Fork, RM 4.7	0	4	0	15	0	1	–	–	–	–	–	–
Indian	0	3	0	2	–	–	–	–	–	–	–	–
Lenz	–	–	–	–	0	17	0	16	0	20	0	17
Neal, middle	–	–	–	–	–	–	–	–	–	–	0	9
Neal, mouth	0	4	1	17	0	16	0	17	0	21	0	17
Neal, upper, above diversion	–	–	–	–	0	17	0	20	0	21	0	17
Neal, upper, below diversion	–	–	–	–	–	–	–	–	0	21	0	12
Rogers	–	–	–	–	–	–	–	–	–	–	–	–
Total	0	17	1	68	0	87	0	97	0	133	0	108

Diazinon	2005		2006		2007		2008		2009		Total	
	d	n	d	n	d	n	d	n	d	n	d	n
Baldwin	2	18	0	3	–	–	–	–	–	–	2	52
Dog	–	–	–	–	–	–	–	–	–	–	0	41
Evans	0	19	0	3	–	–	–	–	–	–	0	83
Hood, East Fork	–	–	–	–	–	–	–	–	–	–	0	32
Hood, Middle Fork	–	–	–	–	–	–	–	–	–	–	0	16
Hood, mouth	0	15	0	7	0	13	0	12	0	11	0	83
Hood, West Fork, mouth	–	–	–	–	–	–	0	6	0	10	0	16
Hood, West Fork, RM 2.5	–	–	–	–	–	–	0	5	0	1	0	6
Hood, West Fork, RM 4.7	–	–	–	–	–	–	–	–	–	–	0	20
Indian	–	–	–	–	–	–	–	–	–	–	0	5
Lenz	0	18	0	3	–	–	0	7	0	14	0	112
Neal, middle	0	16	0	7	0	13	0	12	0	13	0	70
Neal, mouth	0	17	0	7	0	13	0	12	0	14	1	155
Neal, upper, above diversion	0	19	0	7	0	12	–	–	–	–	0	113
Neal, upper, below diversion	0	16	0	7	0	13	0	12	0	14	0	95
Rogers	–	–	–	–	–	–	0	12	0	14	0	26
Total	2	138	0	44	0	64	0	78	0	91	3	925

Table G5. Malathion sample and detection counts, Hood River basin, Oregon, 1999–2009, using data screened at 0.03 micrograms per liter.

[**Abbreviations:** d, number of samples with detections at the reporting level (0.03 micrograms per liter); n, number of samples; RM, river mile; –, not sampled]

Malathion	1999		2000		2001		2002		2003		2004	
	d	n	d	n	d	n	d	n	d	n	d	n
Baldwin	–	–	–	–	–	–	–	–	0	19	0	12
Dog	–	–	–	–	0	11	0	7	0	11	0	12
Evans	–	–	–	–	0	14	0	15	0	20	0	12
Hood, East Fork	0	1	0	12	–	–	0	15	–	–	–	–
Hood, Middle Fork	0	1	0	11	–	–	–	–	–	–	–	–
Hood, mouth	0	3	0	2	0	11	0	7	–	–	–	–
Hood, West Fork, mouth	–	–	–	–	–	–	–	–	–	–	–	–
Hood, West Fork, RM 2.5	–	–	–	–	–	–	–	–	–	–	–	–
Hood, West Fork, RM 4.7	0	3	0	11	0	1	–	–	–	–	–	–
Indian	0	2	–	–	–	–	–	–	–	–	–	–
Lenz	–	–	–	–	0	17	0	15	0	20	1	17
Neal, middle	–	–	–	–	–	–	–	–	–	–	0	9
Neal, mouth	0	3	1	14	1	16	0	17	0	21	1	17
Neal, upper, above diversion	–	–	–	–	0	17	0	20	0	21	0	17
Neal, upper, below diversion	–	–	–	–	–	–	–	–	0	21	0	11
Rogers	–	–	–	–	–	–	–	–	–	–	–	–
Total	0	13	1	50	1	87	0	96	0	133	2	107

Malathion	2005		2006		2007		2008		2009		Total	
	d	n	d	n	d	n	d	n	d	n	d	n
Baldwin	0	18	0	3	–	–	–	–	–	–	0	52
Dog	–	–	–	–	–	–	–	–	–	–	0	41
Evans	0	19	0	3	–	–	–	–	–	–	0	83
Hood, East Fork	–	–	–	–	–	–	–	–	–	–	0	28
Hood, Middle Fork	–	–	–	–	–	–	–	–	–	–	0	12
Hood, mouth	0	15	0	7	0	13	0	10	0	13	0	81
Hood, West Fork, mouth	–	–	–	–	–	–	0	5	0	12	0	17
Hood, West Fork, RM 2.5	–	–	–	–	–	–	0	5	0	1	0	6
Hood, West Fork, RM 4.7	–	–	–	–	–	–	–	–	–	–	0	15
Indian	–	–	–	–	–	–	–	–	–	–	0	2
Lenz	0	18	0	3	–	–	0	7	0	15	1	112
Neal, middle	0	16	0	7	0	13	0	10	0	14	0	69
Neal, mouth	0	17	0	7	0	13	0	10	0	14	3	149
Neal, upper, above diversion	0	19	0	7	0	12	–	–	–	–	0	113
Neal, upper, below diversion	0	16	0	7	0	13	0	10	0	15	0	93
Rogers	–	–	–	–	–	–	0	10	0	15	0	25
Total	0	138	0	44	0	64	0	67	0	99	4	898

Table G6.   Phosmet sample and detection counts, Hood River basin, Oregon, 1999–2009, using data screened at 0.05 micrograms per liter.

[**Abbreviations:** d, number of samples with detections at the reporting level (0.05 micrograms per liter); n, number of samples; RM, river mile; –, not sampled]

Phosmet	1999		2000		2001		2002		2003		2004	
	d	n	d	n	d	n	d	n	d	n	d	n
Baldwin	–	–	–	–	–	–	–	–	0	19	0	12
Dog	–	–	–	–	0	11	0	7	0	11	0	12
Evans	–	–	–	–	0	14	1	15	0	20	0	12
Hood, East Fork	–	–	0	14	–	–	0	15	–	–	–	–
Hood, Middle Fork	–	–	0	15	–	–	–	–	–	–	–	–
Hood, mouth	–	–	0	2	0	11	0	7	–	–	–	–
Hood, West Fork, mouth	–	–	–	–	–	–	–	–	–	–	–	–
Hood, West Fork, RM 2.5	–	–	–	–	–	–	–	–	–	–	–	–
Hood, West Fork, RM 4.7	–	–	0	14	0	1	–	–	–	–	–	–
Indian	–	–	0	2	–	–	–	–	–	–	–	–
Lenz	–	–	–	–	0	17	1	16	2	20	1	17
Neal, middle	–	–	–	–	–	–	–	–	–	–	0	9
Neal, mouth	–	–	0	15	0	16	0	17	0	21	0	17
Neal, upper, above diversion	–	–	–	–	0	17	0	20	0	21	0	17
Neal, upper, below diversion	–	–	–	–	–	–	–	–	0	21	0	11
Rogers	–	–	–	–	–	–	–	–	–	–	–	–
Total	–	–	0	62	0	87	2	97	2	133	1	107

Phosmet	2005		2006		2007		2008		2009		Total	
	d	n	d	n	d	n	d	n	d	n	d	n
Baldwin	0	18	0	3	–	–	–	–	–	–	0	52
Dog	–	–	–	–	–	–	–	–	–	–	0	41
Evans	0	19	0	3	–	–	–	–	–	–	1	83
Hood, East Fork	–	–	–	–	–	–	–	–	–	–	0	29
Hood, Middle Fork	–	–	–	–	–	–	–	–	–	–	0	15
Hood, mouth	0	15	0	7	0	13	0	12	0	12	0	79
Hood, West Fork, mouth	–	–	–	–	–	–	0	5	0	11	0	16
Hood, West Fork, RM 2.5	–	–	–	–	–	–	0	5	0	1	0	6
Hood, West Fork, RM 4.7	–	–	–	–	–	–	–	–	–	–	0	15
Indian	–	–	–	–	–	–	–	–	–	–	0	2
Lenz	3	18	0	3	–	–	1	7	0	12	8	110
Neal, middle	1	16	0	7	0	13	0	12	0	12	1	69
Neal, mouth	0	17	0	7	0	13	0	12	0	12	0	147
Neal, upper, above diversion	0	19	0	7	0	12	–	–	–	–	0	113
Neal, upper, below diversion	0	16	0	7	0	13	0	12	0	12	0	92
Rogers	–	–	–	–	–	–	0	12	0	11	0	23
Total	4	138	0	44	0	64	1	77	0	83	10	892

Table G7. Simazine sample and detection counts, Hood River basin, Oregon, 1999–2009, using data screened at 0.027 micrograms per liter.

[**Abbreviations:** d, number of samples with detections at the reporting level (0.027 micrograms per liter); n, number of samples; RM, river mile; –, not sampled]

Simazine	1999		2000		2001		2002		2003		2004	
	d	n	d	n	d	n	d	n	d	n	d	n
Baldwin	–	–	–	–	–	–	–	–	0	19	0	12
Dog	–	–	–	–	0	11	0	7	0	11	0	12
Evans	–	–	–	–	0	14	0	15	1	20	0	12
Hood, East Fork	0	1	1	14	–	–	0	15	–	–	–	–
Hood, Middle Fork	0	1	0	11	–	–	–	–	–	–	–	–
Hood, mouth	0	3	2	4	0	11	0	7	–	–	–	–
Hood, West Fork, mouth	–	–	–	–	–	–	–	–	–	–	–	–
Hood, West Fork, RM 2.5	–	–	–	–	–	–	–	–	–	–	–	–
Hood, West Fork, RM 4.7	0	3	0	11	0	1	–	–	–	–	–	–
Indian	1	2	–	–	–	–	–	–	–	–	–	–
Lenz	–	–	–	–	8	15	14	16	17	20	8	17
Neal, middle	–	–	–	–	–	–	–	–	–	–	1	9
Neal, mouth	2	3	12	20	5	16	8	17	11	21	7	17
Neal, upper, above diversion	–	–	–	–	0	17	0	20	0	21	0	17
Neal, upper, below diversion	–	–	–	–	–	–	–	–	0	21	0	11
Rogers	–	–	–	–	–	–	–	–	–	–	–	–
Total	3	13	15	60	13	85	22	97	29	133	16	107

Simazine	2005		2006		2007		2008		2009		Total	
	d	n	d	n	d	n	d	n	d	n	d	n
Baldwin	1	18	0	3	–	–	–	–	–	–	1	52
Dog	–	–	–	–	–	–	–	–	–	–	0	41
Evans	0	19	0	3	–	–	–	–	–	–	1	83
Hood, East Fork	–	–	–	–	–	–	–	–	–	–	1	30
Hood, Middle Fork	–	–	–	–	–	–	–	–	–	–	0	12
Hood, mouth	1	15	0	7	0	13	0	10	1	9	4	79
Hood, West Fork, mouth	–	–	–	–	–	–	0	5	0	11	0	16
Hood, West Fork, RM 2.5	–	–	–	–	–	–	0	5	0	1	0	6
Hood, West Fork, RM 4.7	–	–	–	–	–	–	–	–	–	–	0	15
Indian	–	–	–	–	–	–	–	–	–	–	1	2
Lenz	5	18	2	3	–	–	5	7	4	10	63	106
Neal, middle	1	16	0	6	4	13	0	10	0	13	6	67
Neal, mouth	5	17	1	7	4	13	0	10	1	12	56	153
Neal, upper, above diversion	0	19	0	7	0	12	–	–	–	–	0	113
Neal, upper, below diversion	0	16	0	7	0	13	0	10	1	13	1	91
Rogers	–	–	–	–	–	–	0	10	0	11	0	21
Total	13	138	3	43	8	64	5	67	7	80	134	887

Appendix H. Number of Samples Collected at Sites in the Hood River Basin, Oregon, by Site and Year

[Sample size includes all (unscreened) samples. **Abbreviations:** –, no samples collected]

	1999	2000	2001	2002	2003	2004	2005	2006	2007	2008	2009	Total
Baldwin	–	–	–	–	19	12	18	3	–	–	–	52
Dog	–	–	11	7	11	12	–	–	–	–	–	41
Evans	–	–	14	15	20	12	19	3	–	–	–	83
Hood, East Fork	1	16	–	15	–	–	–	–	–	–	–	32
Hood, Middle Fork	1	16	–	–	–	–	–	–	–	–	–	17
Hood, mouth	4	5	11	7	–	–	15	7	13	12	16	90
Hood, West Fork, mouth	–	–	–	–	–	–	–	–	–	6	15	21
Hood, West Fork, RM 2.5	–	–	–	–	–	–	–	–	–	5	1	6
Hood, West Fork, RM 4.7	4	15	1	–	–	–	–	–	–	–	–	20
Indian	3	2	–	–	–	–	–	–	–	–	–	5
Lenz	–	–	17	16	20	17	18	3	–	7	16	114
Neal, middle	–	–	–	–	–	9	16	7	13	12	16	73
Neal, mouth	4	22	16	17	21	17	17	7	13	12	16	162
Neal, upper, above diversion	–	–	17	20	21	17	19	7	12	–	–	113
Neal, upper, below diversion	–	–	–	–	21	12	16	7	13	12	16	97
Rogers	–	–	–	–	–	–	–	–	–	12	15	27
Total	17	76	87	97	133	108	138	44	64	78	111	953

Appendix I. Season of Use for Pesticide Ingredients in the Hood River Basin, Oregon

[**Source:** Hollingsworth, 2009; Peachey, 2009; Pscheidt and Ocamb, 2009; Oregon State University Extension Service, 2010. **Abbreviations:** X, pesticide suitable during the listed season; – pesticide not suitable during the listed season or example product names not provided]

Pesticide	Product names	Spring	Summer	Fall	Winter
Coddling moth mating disruptors					
(Z)-I I-Tetradecen-I-yl acetate	Nomate	X	–	–	–
E-11-tetradecen-1-yl acetate + (e,e)-9,11-tetradecadien-1-yl acetate	Isomate	X	–	–	–
MCPA ester	Checkmate	X	–	–	–
Products for disease control					
1,3 Dichloropropene	Telone II	X	–	X	–
Azoxystrobin	Abound	X	X	–	–
Bicarbonate products	Armicarb, Kaligreen, MilStop, Monterey Bi-Carb	X	X	–	–
Calcium polysulfide	lime sulfur	X	–	X	–
Chloropicrin	–	–	–	X	–
Dazomet	Basamid G	X	–	–	–
Dichloran	Botran	X	–	–	–
Dimethyl phenol	Gallex	–	–	–	–
Iprodione	Iprodione	X	–	–	–
Kaolin	Surround	X	–	–	–
Mancozeb	–	X	–	–	X
Metam sodium	Vapam, Sectagon 42, Metam CLR	X	–	X	–
Methyl bromide	–	–	–	X	–
Methyl phenol	Gallex	–	–	–	–
Mono- and dipotassium salts of phosphorous acid	Agri-Fos	X	X	–	–
Monopotassium phosphite + dipotassium phosphite	Fosphite	X	X	–	–
Sulfur products[1]	–	X	X	X	–
Fungicides					
Boscalid	Endura, Pristine	X	X	X	–
Captan	Captan, Captec	X	X	X	–
Chlorothalonil	Bravo Weather Stik	X	–	–	–
Copper products	–	X	–	X	–
Cyprodinil	Vangard, Switch	X	X	–	–
Dodine	Syllit	X	–	–	–
Fenarimol	Rubigan	X	X	–	–
Fenbuconazole	Indar	X	X	–	–
Fenhexamid	Elevate, CaptEvate	X	X	X	–
Fludioxinil	Switch	X	X	–	–
Fosetyl-aluminum	Aliette	X	X	X	–
Kresoxim-methyl	Sovarn	X	X	–	–
Metalaxyl	Ridomil Gold	X	X	X	–
Metconazole	Quash	X	X	–	–
Myclobutanil	–	X	X	–	–
Potassium bicarbonate	Remedy	–	–	–	–
Propiconazole	Tilt	X	X	–	–
Pyraclostrobin	–	X	X	X	–
Pyrimethanil	Scala	X	–	–	–
Quinoxyfen	Quintec	X	X	–	–
Sodium, potassium, and ammonium phosphites	Phostrol	X	X	–	–
Sodium tatrathiocarbonate[1]	Enzone	X	–	–	–
Streptomycin	Agrimycin	X	–	–	–

Appendix I.   Season of use for pesticide ingredients in the Hood River basin, Oregon.—Continued

[**Source:** Hollingsworth, 2009; Peachey, 2009; Pscheidt and Ocamb, 2009; Oregon State University Extension Service, 2010. **Abbreviations:** X, pesticide suitable during the listed season; – pesticide not suitable during the listed season or example product names not provided]

Pesticide	Product names	Spring	Summer	Fall	Winter
Fungicides—Continued					
Tebuconazole	Elite, Orius	X	X	–	–
Terramycin	Mycoshield	X	–	–	–
Thiophanate-methyl	–	X	X	X	–
Trifloxystrobin	–	X	X	X	–
Triflumizole	–	X	X	–	–
Triforine	Funginex	–	–	–	–
Ziram	–	X	X	X	–
Products to control fruit drop					
Aminoethoxyvinylglycine hydrochloride	Retain	X	X	–	–
Napthalene acetic acid (NAA)	NAA	–	X	–	–
Herbicides					
2,4-D	Crossbow, Curtail, Weedmaster, Pasturemaker, Cimarron Max	X	X	X	–
2,4-D amine	Saber, Weed-Rhap A4d, Dri-Clean Herbicide	X	–	–	–
2,4-D ester	Crossbow	X	–	–	–
Aminopyralid	Milestone	X	X	X	–
Atrazine	–	X	–	–	–
Bromacil	Krovar	X	–	–	–
Carfentrazone	Aim	X	–	–	–
Chlorsulfuron	Telar	X	–	X	–
Clethodim	Envoy, Prism, Select	–	–	–	–
Clopyralid	–	X	X	X	X
Dicamba	Banvel, Vanquish, Clarity, Weedmaster, Pasturemaker, Latigo	X	X	–	–
Dichlobenil	Casoron	–	–	–	X
Diquat	Reglone	–	–	–	–
Diuron	–	X	X	X	X
Fluazifop	Flusilade	–	–	–	–
Flumioxazin	Chateau	–	–	X	–
Fluroxypyr	Starane, Surmount, PastureGard	X	–	–	–
Glufosinate ammonium	Rely	–	–	–	–
Glyphosate	–	X	X	X	X
Halosulfuron	Sandea	–	–	–	–
Hexazinone	–	X	X	X	X
Imazapic	Plateau	–	X	X	–
Imazapyr	–	–	X	–	–
Isoxaben	Gallery or Gallery T&V, Showcase, Snapshot	X	X	X	–
MCPA	–	X	–	X	–
Mesotrione	Callisto	X	–	–	–
Metsulfuron methyl	Cimarron Max, Escort	X	X	X	–
Napropamide	Devrinol	X	–	X	X
Norflurazon	Solicam	X	–	X	X
Oryzalin	–	X	–	X	–
Oxyfluorfen	–	X	–	X	X
Paraquat	Gramoxone Inteon, Firestorm, Cyclone	X	–	X	X
Pendimethalin	Prowl H2	X	–	X	X

Appendix I.   Season of use for pesticide ingredients in the Hood River basin, Oregon.—Continued

[**Source:** Hollingsworth, 2009; Peachey, 2009; Pscheidt and Ocamb, 2009; Oregon State University Extension Service, 2010. **Abbreviations:** X, pesticide suitable during the listed season; – pesticide not suitable during the listed season or example product names not provided]

Pesticide	Product names	Spring	Summer	Fall	Winter
Herbicides (Continued)					
Picloram	–	X	X	X	X
Pronamide	–	–	–	X	X
Rimsulfuron	Matrix	X	–	–	–
Sethoxydim	Poast	–	–	–	–
Simazine	–	X	–	X	X
Sulfometuron methyl	–	–	–	X	–
Tebuthiuron	Spike	X	X	X	X
Terbacil	Sinbar	X	–	X	–
Triasulfuron	Amber	–	–	–	–
Triclopyr	–	X	X	–	–
Triclopyr ester	–	X	X	–	–
Trifluralin	Showcase, Snapshot, Treflan	–	–	–	X
Insecticides					
Abamectin	–	X	X	–	–
Acephate	–	–	–	–	–
Acetamiprid	Assail	X	X	–	–
Azadirachtin	Azatin, Neemix	X	X	–	–
Azinphos methyl	Guthion	X	X	–	–
Bifenazate	Acramite	–	X	–	–
Bifenthrin	Brigade	X	X	–	–
Buprofezin	Applaud, Centaur	X	X	–	–
Carbaryl	Sevin	X	X	–	–
Chlorantraniliprole	Voliam Flexi	X	X	–	–
Chlorpyrifos	–	X	–	–	X
Clothianidin	–	X	X	–	–
Cyfluthrin	Baythroid	–	–	–	–
Deltamethrin	–	X	X	–	–
Diazinon	Diazinon	X	X	–	–
Dicofol	Kelthane	–	X	–	–
Diflubenzuron	Dimilin	–	–	–	–
Dimethoate	Dimethoate	X	X	–	–
Disulfoton	–	–	–	–	–
Dormant oil	–	–	–	–	–
Emamectin benzoate	Proclaim	X	X	–	–
Endosulfan	–	X	X	–	–
Esfenvalerate	Asana	X	–	–	–
Fenbutatin oxide	Vendex	X	X	–	–
Fenpropathrin	–	X	–	–	X
Gamma-cyhalothrin	–	X	–	–	X
Imidacloprid	–	X	X	–	–
Indoxacarb	Avaunt	X	X	–	–
Insecticidal soap	M-Pede, others	X	X	–	–
Iron phosphate	–	–	–	–	–
Lambda-cyhalothrin	–	X	–	–	X
Malathion	Malathion	X	X	–	–
Metaldehyde	–	–	–	–	–
Methidathion	Supracide	X	–	–	–
Methomyl	Lannate	X	–	–	–
Methoxyfenozide	Intrepid	X	X	–	–
Methyl parathion	Methyl 4EC	–	–	–	–
Novaluron	Rimon	X	X	–	–

Appendix I.   Season of use for pesticide ingredients in the Hood River basin, Oregon.—Continued

[**Source:** Hollingsworth, 2009; Peachey, 2009; Pscheidt and Ocamb, 2009; Oregon State University Extension Service, 2010. **Abbreviations:** X, pesticide suitable during the listed season; – pesticide not suitable during the listed season or example product names not provided]

Pesticide	Product names	Spring	Summer	Fall	Winter
Insecticides (Continued)					
Oxamyl	Vydate	X	X	–	–
Permethrin	–	X	–	–	X
Petroleum or paraffinic oil	Horticultural mineral oil	X	X	X	–
Phosmet	Imidan	X	X	–	–
Pyrethrins/pyrethrum	–	–	–	–	–
Pyriproxyfen	–	X	X	–	X
Rotenone	–	–	–	–	–
Rynaxypyr	–	X	X	–	–
Spinetoram	–	X	X	–	–
Spinosad	Entrust, Success	X	X	–	–
Spirodiclofen	Envidor	X	X	–	–
Spirotetramat	–	X	X	–	–
Tebufenozide	Confirm	–	–	–	–
Thiacloprid	–	X	X	–	–
Thiamethoxam	Axtara, Platinum, Voliam Flexi, Actara	X	X	–	–
Zeta cypermethrin	Mustang Max	–	–	–	–
Miticides					
Acequinocyl	Kanemite	X	X	–	–
Bifenzate	Acramite	X	X	–	–
Clofentezine	Apollo	X	X	–	–
Etoxazole	Zeal	X	X	–	–
Fenpyroximate	Fujimite	X	–	–	–
Formetanate hydrochloride	Carzol	X	–	–	–
Hexythiazox	Onager, Savey	X	X	–	–
Propargite	Omite	–	X	–	–
Pyridaben	Nexter	X	X	–	–

[1]Also used as an insecticide

Appendix J. Pesticide Products Known to be Used in the Hood River Basin, Oregon, but not Analyzed for this Study

[**Source:** John Buckley, East Fork Irrigation District, oral commun., 2010; Steve Castagnoli, Oregon State University Extension Service, oral commun., 2010; Nate Lain, Hood River County Weed and Pest Division, oral commun., 2010; Brian Walker, Oregon Department of Transportation, oral commun., 2010]

Pesticide	Product names	Known use in the Hood River basin
Products used for disease control		
Mancozeb	Dithane, Mancozeb	Commonly used in orchards (Feb/Mar)
Sulfur products	Thiolux, Microthiol Disperss, Kumulus	Commonly used in orchards (Feb/Mar or Sept/Oct)
Fungicides		
Myclobutanil	Rally, Spectracide Immunox	Commonly used in orchards
Thiophanate-methyl	Topsin M, Halt	Orchard use is expected to increase (Sept/Oct)
Trifloxystrobin	Flint, Gem	Commonly used in orchards
Triflumizole	Procure	Commonly used in orchards
Ziram	Ziram	Commonly used in orchards (Sept/Oct)
Herbicides		
Clopyralid	Stinger, Transline, Curtail, Redeem R&P	Commonly used in forests
Glyphosate	Roundup	Commonly used in orchards and forests, along canals, roads, and railroads
Metsulfuron methyl		Commonly used in forests and along railroads
Oryzalin	Surflan	Commonly used in orchards
Oxyfluorfen	Goal, Showcase	Commonly used in orchards
Sulfometuron methyl	Oust	Commonly used in forests
Triclopyr ester	Remedy	Commonly used in forests
Insecticides		
Abamectin	Agri-Mek	Common use in orchards (late Apr - early Jun)
Acetamiprid	Assail	Neonicotinoid - class expected to be more common in orchards
Clothianidin	Clutch	Neonicotinoid - class expected to be more common in orchards
Deltamethrin	Battalion	Pyrethroid - class commonly used in orchards (Feb-Mar)
Fenpropathrin	Danitol	Pyrethroid - class commonly used in orchards (Feb-Mar)
Gamma-cyhalothrin	Proaxis	Pyrethroid - class commonly used in orchards (Feb-Mar)
Lambda-cyhalothrin	Warrior, Warrior II	Pyrethroid - class commonly used in orchards (Feb-Mar)
Rynaxypyr	Altacor	New product likely to have widespread use in orchards
Spinetoram	Delegate	New product likely to have widespread use in orchards
Thiacloprid	Calypso	Neonicotinoid - class expected to be more common in orchards

www.ingramcontent.com/pod-product-compliance
Lightning Source LLC
Chambersburg PA
CBHW081551170526
45166CB00009B/2664